The Evolution of Gun Making

This book is dedicated to a long-standing friend, Robert (Bob) Gordon, Professor Emeritus at Yale University, whom I first met when he visited the Industrial Museum in Sheffield where I was curator in 1984. Even at first encounter we soon began discussing gunmaking by machinery and interchangeability and it was clear that he was one of the early historians and pioneers in this field of industrial archaeology. It was his encouragement which led me down this tortuous and mystery-laden path of discovery occupying over half a lifetime and for which I am sincerely grateful.

The Evolution of Gun Making

Machine made weapons, 1700–1820

PETER G. SMITHURST

Pen & Sword
MILITARY
AN IMPRINT OF PEN & SWORD BOOKS LTD.
YORKSHIRE – PHILADELPHIA

First published in Great Britain in 2024 by
PEN AND SWORD MILITARY
An imprint of
Pen & Sword Books Limited
Yorkshire – Philadelphia

ISBN 978 1 03610 778 9

A CIP catalogue record for this book is available from the British Library.

Typeset in Times New Roman 10/12 by
SJmagic DESIGN SERVICES, India.
Printed and bound in the UK by CPI Group (UK) Ltd.

Pen & Sword Books Limited incorporates the imprints of Atlas, Archaeology, Aviation, Discovery, Family History, Fiction, History, Maritime, Military, Military Classics, Politics, Select, Transport, True Crime, Air World, Frontline Publishing, Leo Cooper, Remember When, Seaforth Publishing, The Praetorian Press, Wharncliffe Local History, Wharncliffe Transport, Wharncliffe True Crime and White Owl.

For a complete list of Pen & Sword titles please contact
PEN & SWORD BOOKS LIMITED
George House, Units 12 & 13, Beevor Street, Off Pontefract Road,
Barnsley, South Yorkshire, S71 1HN, England
E-mail: enquiries@pen-and-sword.co.uk
Website: www.pen-and-sword.co.uk

Or
PEN AND SWORD BOOKS
1950 Lawrence Rd, Havertown, PA 19083, USA
E-mail: Uspen-and-sword@casematepublishers.com
Website: www.penandswordbooks.com

MIX
Paper | Supporting responsible forestry
FSC® C013604

Contents

Preface

Many conceive the 'Industrial Revolution' to have been a uniquely British thing, and undoubtedly Britain was pre-eminent in many fields of endeavour. The developing textile industries were replete with such names as Kay's 'flying shuttle', Crompton's 'mule', Arkwright's 'spinning jenny' and 'water frame', each contributing a step in the process of transforming a cottage industry into a factory system. Often such steps in mechanisation were seen not as 'labour saving' but as 'job destroying' and in the wake of such innovations were the machine-destroying 'Luddite riots'.

When gathered into factories, such machines needed a source of power far beyond that provided by the human touch and relied initially upon waterpower, so factories were restricted to river-banks. In 1698 Thomas Savery invented a steam-powered pump, which helped to overcome the problem of flooding in shallow coal mines and could have been used to convey water from a river to a remote water wheel. Fourteen years later, Thomas Newcomen invented a more efficient engine which was used extensively across Britain and Europe, again for pumping out mines. This was a spur to the coal and other mining industries but had no impact on factories. James Watt made significant improvements to steam pumping engines but his major contribution to steam engine development came in 1781 when he invented the rotary engine. Replacing the waterwheel with a Watt rotary engine transformed the factory, giving it more power, available in all seasons, and removing the need for it to be built alongside a river.

Britain also made significant contributions to the manufacture of iron and steel, in particular Benjamin Huntsman's process developed around 1742, in which for the first time in Europe, steel was produced in a molten state, and that alone laid the foundations of the modern steel industry. Cast iron began to be used for civil engineering projects, notably the world famous cast-iron bridge at Ironbridge Gorge. In the later 18th century Henry Maudslay pioneered precision manufacturing with his screwcutting lathe, allowing truly accurate threads to be made for the first time. He went on to build machines for Bramah's lock making venture and Brunel's block making machines, both milestones in mechanised manufacturing. Much more could be said but that little foray gives some idea of Britain's contribution to 'the Industrial Revolution'.

However, another less dramatic but equally important revolution was taking place in the manufacture of military firearms. In France, the pursuit of interchangeability – the ability to exchange identical parts within identical mechanisms without modification or 'adjustment' – led to a new industrial revolution in which traditional hand-craft skills were replicated by machines. This mechanisation was further pursued in early 19th century Russia and Britain. At this time, it began to be pursued with great zeal in America where many new machines were devised for undertaking nearly all aspects of the work. This has become known as 'The American System of Manufacture' though, as will be seen, the system evolved through various times in different places.

Much has been written on the topic, particularly in the second half of the 20th century, which examine the political, military, and economic factors which drove it. Industrial, technology and firearms historians may pay 'lip service' to these gunmaking technologies but few, if any, examine the technology itself and those which do tend to give an overall picture as opposed to providing a technical narrative. Yet these technologies have greater importance than simply for making guns. They were adopted and adapted to make other, often complex and expensive, goods at affordable prices and provided the foundations of the 'consumer society' which forms part of our everyday lives.

This study, therefore, whilst following the historical progression, will focus on the technology with detailed analyses of the functional characteristics of the machines and what they accomplished – or, in some cases, didn't.

This first volume examines in detail the manufacture of the French Model 1777 and Russian Model 1808 muskets; the second will focus on the Enfield Pattern 1853 rifle. A variety of fortuitous considerations surround the choice of these weapons. The first is that they are the only ones within this period whose manufacture is covered by contemporary accounts. These accounts provide varying levels of detail and where it has been felt necessary, or just simply desirable, any shortfalls have been made up for by what is hoped to be logical deduction and explanatory diagrams.

The second, and possibly more important consideration, is that these weapons are interlinked by structural parallels with the French Model 1777 musket and bayonet.

As with any research, and especially research of a historical nature, we are limited by the evidence that is available and it can never be guaranteed that it is all the evidence. I began research on the manufacture of the Enfield Pattern 1853 rifle many years ago, before the internet existed, and finding relevant material was very much a matter of luck, stumbling across a useful reference or a publication while rummaging in a bookshop or library. While the internet has opened up so much, one can only search for something specifically if it is known to have existed, but the chances of success are certainly multiplied by the use of 'keywords'. But even then, there is no guarantee that one has every word that was ever written about something. I am reminded of an apt quotation by Cardinal John Henry Newman (1801–1890) which appeared on the flyleaf of one of my graduate chemistry textbooks: "*A man would do nothing if he waited until he could do it so well that no one could find fault*". I am also reminded of a quotation by Thomas Jefferson: *"He who knows nothing is closer to the truth than he whose mind is filled with falsehoods and errors"*. What follows is therefore based on the evidence available at the time of writing and I take full responsibility for any faulty deductions, factual errors, and omissions. If anyone can show me to be undeniably wrong in any of these, I would be grateful to hear from them. Learning is a lifelong process and only by acknowledging our errors or misconceptions can we improve our knowledge and understanding.

Acknowledgements

While this work has been very much a 'labour of love', based on a fascination with the topic of historical arms manufacturing technology, it would not have been possible without the aid and support of others for which unbounded thanks are offered.

In respect of this, the first volume, I have to give thanks especially to Royal Armouries and colleagues there. My former role as a Senior Curator of Firearms, and latterly as Curator Emeritus, have given me unrestricted access to the collections which has been vital. Particular thanks are due to:

Laura Bell, Director of Collections;
Graeme Rimer, formerly Academic Director and now also Curator Emeritus;
Jonathan Ferguson, Keeper of Firearms and Artillery;
Lisa Traynor, Curator of Firearms;
Stuart Ivinson, Librarian;
Philip Abbott, Archivist;
Mark Bennet, Director of Research;
Caroline Moran, Image Licensing;
all of whom have given unstinting help.

Special thanks for free use of images of Royal Armouries Collections items.

Thanks are also due to:

Alex McKenzie, Curator at Springfield Armory National Historic Site.

Agathe Formery of the Musée de l'Armée and Gladys Pilastrini, Barbara Van Kets, and Florence Hemici of Agence photographique de la RMN, Paris, for details and images of Blanc's gauges and guiding me through the purchase and licensing processes.

Nathalie Naudí, Agence photographique, Musée des Arts et Métiers - Le CNAM, Paris, for images and guiding me through the purchase and licensing processes.

Thanks are also due to Ken Alder for providing a copy of a now lost drawing showing lock components of the French 1777 Musket. And last but not least my grateful thanks to the publishers, especially Tara Moran and Harriet Fielding, for their acceptance of this work and their guidance through the intricacies of getting it into print.

Chapter 1

The Manufacture of the French Model 1777 and Russian Model 1808 Muskets

Introduction

It may seem strange that these two muskets have been grouped together for this study but, as in Kipling's ditty, 'The Ladies', *"The Colonel's lady and Judy O'Grady are sisters under their skin"*,[1] these are indeed sisters under their skins, if not exactly twins. Foremost in this assessment is that their similarities are too numerous to be coincidental – it is inconceivable that two different and remote designers could have evolved so many conceptually identical features. Added to that is the fact that the Russian military hierarchy were aware of the 1777 musket's design through a publication by Herman Cotty in 1806 and which no doubt stimulated their interest.[2]

A further consideration is that these two weapons are the subjects of the earliest known publications which describe military firearms manufacture in detail. For the French M.1777 we have Cotty's *Mémoire sur la fabrication des armes portatives de guerre* [Memorandum on the manufacture of portable weapons of war] of 1806 and Dale's *Memoire sur la mise en bois ou montage du fusil, la trempe de la platine, la visite et la recette de l'arme finie* [Memorandum on stocking, hardening and tempering the lock, inspection and acceptance of the finished weapon] of 1810, with

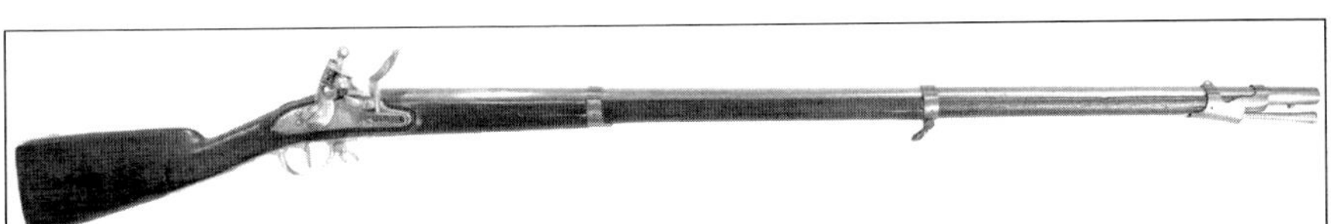

Fig. 1.1. French Model 1777 musket made at Versailles (XII.2231, © Royal Armouries)

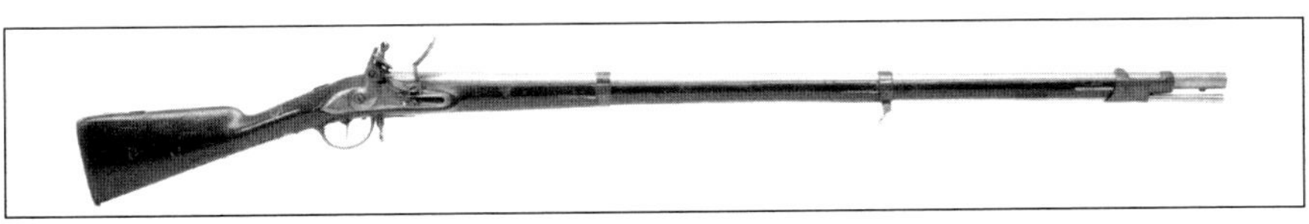

Fig. 1.2. Russian M.1808 Musket made at Tula (XII.2351, © Royal Armouries)

earlier technologies being described and illustrated by Diderot in various editions of his Encyclopaedia; for the Russian M.1808 we have Gamel's *Description of the Tula Weapon Factory in regard to historical and technical aspects,* published in 1826 with a translation appearing in 1988. If these are compared with practices detailed by two Portuguese gunmakers, Fiosconi and Guserio, in *Espingarda Perfeyta* in 1718, the earliest encountered treatise on gunmaking, and those to be encountered in Birmingham and Enfield in the 1850s, the lineage is clearly to be seen. Some are virtually unchanged. It will be shown, however, that the French and Russian muskets embrace a period in which military gunmaking underwent significant technological development. Tentative steps in this direction were being taken in France in the 1770s while more extensive 'mechanisation' was adopted and developed in Russia in the early decades of the 19th century. Similar advances were, of course, being made in America during this period and whilst important, they largely fall outside the timeframe of this particular study but will be dealt with in a second volume.

It is doubtful whether any of these French and Russian publications have been studied in any detail since they were first published. Apart from anything else, they highlight the evolution of the technologies employed to accomplish their goals.

The French M.1777 musket and its designer, Honoré Blanc, occupy a special place in the history of military firearms. Honoré Blanc is one of the unsung heroes of arms design and manufacture. It would seem from his own testimony[3] that he first entered the sphere of arms manufacture in 1748 and, having spent some time as Master Armourer at the Charleville factory,[4] in 1763 he was appointed *côntroleur de la manufacture d'armes* at St. Etienne. There he designed and created all the firearms issued in 1764, 1766, 1768, 1774 and culminating in what became his iconic *Modèle 1777 Fusil d'infanterie* musket and bayonet. These not only became the standard patterns on which French military arms were based but they, or elements of them, and their manufacture had a far-reaching impact in Europe, Russia, America, and England well into the 19th century.

At the same time as he designed the musket, he also designed the *Modèle 1777 Pistolet de Cavalerie*, Fig. 1.3, which incorporated his sloping pan, characteristic

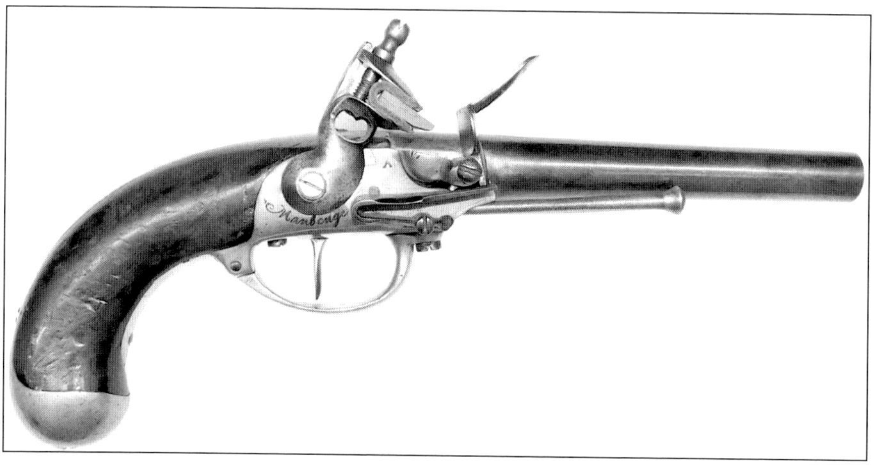

Fig. 1.3. The Model 1777 cavalry pistol. (Author's collection)

of derivatives of his 1777 musket, plus other innovative design and construction features. Most notable amongst these is what might be referred to as a 'monobloc' system comprising a single casting which was closely fitted to the barrel and housed the lock mechanism and the butt; in many respects resembling a large, side-hammer, 'boxlock' pistol.

In 1776 General Gribeauval became Inspector General of Artillery in France and began his reforms, part of which was a redesign of artillery and carriages and the pursuit of standardisation. The manufacture of any military arms in France at this time came under the control of the '*Artillerie*', and no doubt Gribeauval became aware of Blanc's work through his 1777 Musket. As a result of this, in the following year, 1778, Blanc was appointed Principal Controller of the other three arms manufactories in France - Charleville, Maubeuge and Tulle - in addition to St. Etienne.

After visiting these manufactories in 1782, his attention was drawn to *the multiplicity of imperfections which negligence had left in the manufacture of arms* and he *gave himself up to every reflection capable of suggesting to him the means of remedying it.*[5] One of his first actions was to implement standardisation by specifying the proportions and dimensions of the various weapons and their components. He created a plan of work which embraced all the parts of the weapon and

> *. . . occupied himself with the invention of various measures, tools, machines and dies, hitherto unknown in our manufactures, in considerable number to facilitate the work of more than 800 workers, and to give to this work all the precision possible.*[6]

This embodies the concept and principles of interchangeable manufacture and constitutes a major change in the approach to manufacturing a composite and geometrically complex product. It assumes even greater importance when one considers that some workers, such as the lock makers, often worked from their homes at some distance from the manufactory itself, and the only means of ensuring their work exhibited uniformity was using matching dies, jigs and gauges. It was probably around this time that Blanc produced the earliest known comprehensive set of inspection gauges to ensure this uniformity, Fig. 1.4.

In or before 1786, Blanc's work had come to the notice of someone he refers to as *M. Châtelet* but who was likely to have been the *Duc du Châtelet*, a prominent member of the Comité Militaire, and who, like Gribeauval, was impressed with Blanc's work and aspirations. He intervened with the Minister who gave Blanc space at Vincennes to develop his ideas. Here he devoted his efforts to the manufacture of locks, resulting in two documents which are of primary importance to the study of both the practical aspects of lock-making and the early pursuit of interchangeability. The first is his own *mèmoire* in which he presented to the National Assembly his report, particularly on the manufacture of gun locks with interchangeable parts and their utility:

> *. . . that the pieces composing a lock being manufactured separately, may, after they have been completed, be intermingled and confused*

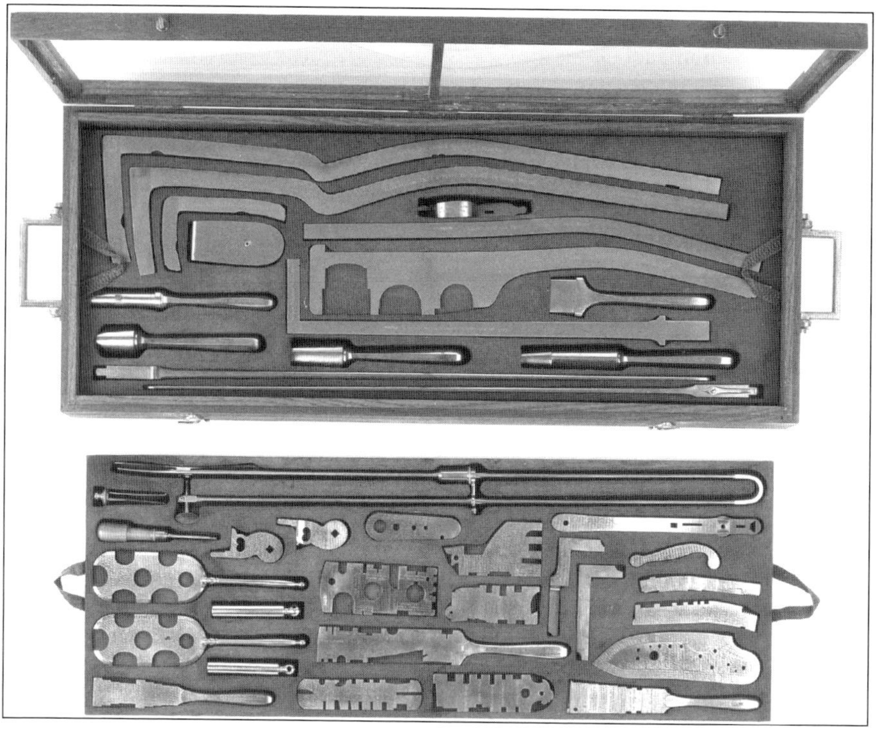

Fig. 1.4. A set of gauges for the M. 1777 musket, made by Blanc c.1777-1795, in the Musée de l'Armée in Paris. (Inv. No. 2013.0.11. Photo (C) Paris - Musée de l'Armée, Dist. RMN-Grand Palais - © Pascal Segrette)

> *together, and then each taken at random to serve to form the lock, by means of which there is no-one who is not able to reassemble a lock, with the same ease as if it were only a question of changing the flint.*

and

> *. . . one would enjoy the inestimable advantage of carrying in a crate, spare parts of all kinds, hardened and tempered, polished, and ready to replace a broken or missing piece, since it would no longer be necessary to have an armourer in charge provided that they* [the soldier] *can dismantle them, which can easily be taught.*[7]

The second is a report by Messrs le Roy, de la Place, Borda, and Coulomb of the Academie Royale des Sciences on Blanc's submission to the National Assembly and which endorsed his claims in no uncertain terms.[8] Blanc sent another petition to the National Assembly which resulted in the War Office appointing a commission of artillery officers to investigate his claims and which echoed the Academy's findings.[9]

This was on the eve of great upheavals in which the ebb and flow of political ideals in France had an impact on the way arms were produced, and these interactions are covered in excellent detail in Alder's book, *Engineering the Revolution.*

Blanc, realising that he would not get support from the state, asked to be released from his official post and to be allowed to establish himself as a private entrepreneur with an interest-free loan of 30,000 *livres*. The commissioners concurred, the government reserving the right to make copies of Blanc's machines, and the Minister approved the project on 3rd April 1792. In November 1793, while the Committee was hastily assembling the much-vaunted Paris Manufactory, he was authorised to purchase the Ursuline convent in Roanne, along with a water mill. In return, Blanc was to manufacture 30,000 interchangeable locks per year.[10]

So how did Blanc make his interchangeable locks? His *mèmoire* is less informative on such details than we might hope and is more concerned with presenting an overall picture of the then state of arms production in France and what his proposals would accomplish. He claims that he, *invented for this purpose several machines and a large number of tools of such usefulness,*[11] but unfortunately he does not elaborate on these.

On the other hand, the Academy was more intent on establishing whether Blanc's claims in relation to the production of locks with interchangeable components were accurate and to establish whether the system he was proposing was viable. Their report does contain a little more detail of some manufacturing processes but still falls far short of describing them all.

Fortunately, many of these details are collectively provided in three other, more or less contemporary, accounts mentioned earlier.

Equally, the Russian Model 1808 musket occupies a special place in the history of firearms manufacture. Gamel, in his *Description of the Tula Weapon Factory in regard to historical and technical aspects,* published in 1826, notes that interchangeability was being pursued. In 1817 John Jones and his son, John, were recruited from Birmingham to Tula. Jones introduced numerous improved methods of manufacture, in particular die-stamping of lock components which enabled interchangeability to be ostensibly achieved. He also developed *'extremely useful and complicated equipment for the production of the gunstock and the lock screws'* but, unfortunately, Gamel does not elaborate on the 'stocking equipment'. What is abundantly clear from Gamel's book is that a far greater number of sophisticated machines were being employed than in France, and possibly than in America, the widely accepted source of mechanised gunmaking technology along with interchangeability.

Through a combination of these documents, drawing upon other scattered sources and the creation of diagrams to elucidate various operations, it is possible to arrive at a greater understanding of some of the manufacturing processes being developed in this period.

While it would be satisfying to treat these in accordance with the popular phrase, *lock, stock and barrel* it has been decided, to allow for easier cross-referencing, to follow Cotty's and Gamel's order of the barrel followed by the lock and lastly the stock, of which Gamel, like Cotty, has little to say.

Chapter 2

Barrel Manufacture

2.1 Introduction and the properties of wrought iron

Wearisome work it is to experience the heat of the fire, but worse it were for men to be shy of this trial, to remain unskilled in this office. Without mortification little is achieved, and without resolution nothing is gained. I speak to those who, when starting on this road lack determination because they are impatient, and allow weariness to be an obstacle to them without caring for better work, and to those who, overcoming such doubts, strive by all means to be a good craftsman, in order to become a better Master. To these we repeat, that the utmost perfection and excellence of works is to repeat corrections, even though this may be irksome, for otherwise they will become negligent and so little practised, that they will call difficulties what are the effects of laziness, which has never occasioned anything convenient, nor improved in his craft any man born of this Monster.[1]

In the comment above the writers of *Espingarda Perfeyta* in the early 18th century acknowledge that the making of a gun barrel is both very skilled and very arduous work. It also expresses what can be seen as the creed of the true craftsman - the constant seeking for perfection in what they produce. The authors might also have added, had they been aware of it, a very enlightened commentary made fifty years previously and attributed to Glanvill:[2] *Iron seemeth a simple metal, but in its nature are many mysteries, and men who bend to them their minds shall, in arriving days, gather therefrom great profit, not to themselves alone, but to all mankind.*

As will become apparent, both are very pertinent considerations in the creation of gun barrels. It has already been noted that the common feature of any gun is a barrel and by the time *Espingarda Perfeyta* was written, for hand-held weapons wrought iron had been the preferred material for several centuries, *for only iron can resist the adversity of the materials of which powder is composed, especially sulphur, which united with fire consumes metals and none better than iron resists it.*[3]

Wrought iron also has greater strength than the other possible material, bronze, allowing it to resist a greater internal force from the explosion of gunpowder,[4] enabling the creation of lighter and more powerful guns. Equally, it may have rested on its ready availability and, therefore, its lower initial cost. The choice was unlikely, however, to have been based on ease of manufacture since the making and working of wrought iron are lengthy processes requiring high degrees of skill and stamina at every stage.

Prior to the advent of gunpowder weapons, no evidence has been found for the existence of, or the need for, iron tubes of any type, cast or wrought. Water, if it was piped at all, was carried in wooden pipes made from hollowed out elm tree trunks and which have a history dating to the Roman era, if not before. The ancient Greeks had used both lead and bronze for domestic water supplies. The ancient Egyptians had even used copper pipes.[5] The only requirement for light, wrought iron tubes, other than for use in firearms, does not seem to have arisen until the early part of the 19th century when gas lighting began to be introduced. Even then we are told that in 1815, William Murdoch, the pioneer of gas lighting, when installing a gas plant at the Soho Works near Birmingham, used old musket barrels screwed together to pipe gas throughout the works.[6] The first gunmakers, therefore, were confronted with a completely new requirement and had to develop a new manufacturing technology to satisfy it.

We can only surmise what methods were used prior to the early 18th century, at which point the first detailed account of making the barrels of firearms appears. It is imagined that by the time the need for forged gun barrels had arisen, ironworkers were familiar with the fibrous-like nature of wrought iron in which the body of iron was interspersed with streaks of slag or other inclusions to create a non-homogenous metal. As a lump of wrought iron was drawn out into a bar, so too these inclusions would be drawn out in the same direction. They created lines of weakness and while a bar of iron remained strong along its length, it was weaker across its width. In this sense, it can be likened to a piece of wood. This 'weakness' is clearly demonstrated in a sample prepared by Henry Bessemer who hammered a bar of wrought iron which, even at dark red heat, disintegrated, Fig. 2.1.[7]

This image reinforces the fact that wrought iron can only be forged when at high, 'sparkling', welding temperature. It should also be noted that at this time this anisotropic nature was also shared by steel since it was made by the Cementation process, which involved baking bars of wrought iron in beds of charcoal at red-heat for several days. It was, in effect, a large-scale case-hardening process and produced the so-called 'blister steel'. There was no melting involved and so the physical structure of the iron was retained by the steel. Only around 1742 did Benjamin Huntsman in Sheffield succeed in producing molten steel for the first time in Europe. By melting blister steel in crucibles, it was possible to remove the impurities as a 'slag', which floated to the surface and could be skimmed off, creating for the first time a homogenous steel which had a fine crystalline structure, as opposed to the coarser, fibrous grain possessed by wrought iron and blister steel.

By the early 18th century, sophisticated techniques had evolved for the making of forged and welded barrels, leading to two basic types. On the one hand there are those which can be described as 'plain' barrels and on the other, those that are

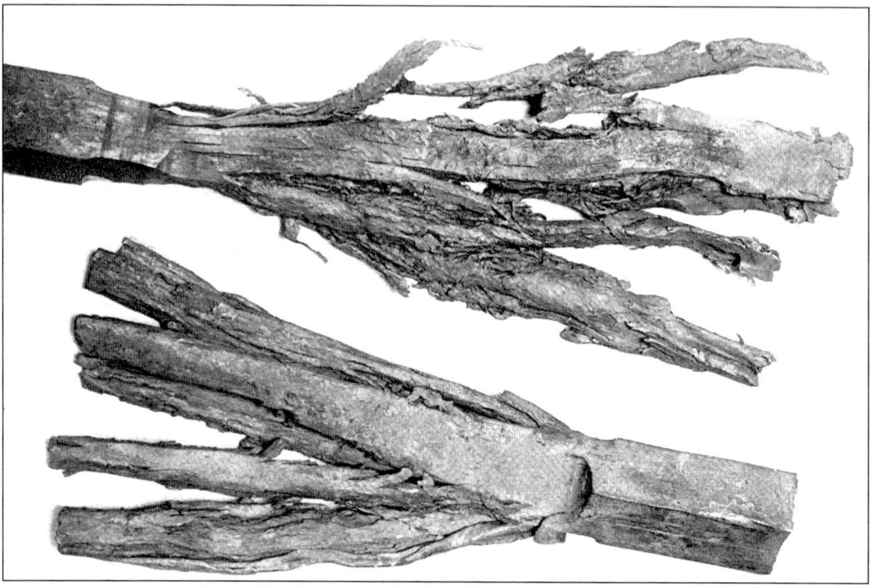

Fig. 2.1. A bar of wrought iron hammered below correct forging temperature reveals its 'fibrous' nature. (Bessemer, 1905, pl. XXIV)

loosely referred to as 'twist' or 'Damascus' barrels. Initially, at least, Damascus barrels having the 'grain' running circumferentially instead of longitudinally as in the plain barrels, were created for their supposedly superior strength, demonstrating a clear understanding of the physical properties and nature of wrought iron. However, as shown later, Greener argued that this was a myth!

Reflection on the problems of making a welded iron tube perhaps suggests that 'plain' barrels came first and then, after a greater understanding and appreciation of the anisotropic properties of iron, and the earlier form of steel, had been acquired, a method was found of using it to what was thought to be best advantage in the twist barrel.

These 'fibrous' properties of iron are very well illustrated by two specimens. In Fig. 2.2 is shown part of a twist or Damascus barrel which has failed and has separated at a helical welded joint between two strands of the strip or 'ribband' of iron from which it was made. It has failed by separating across the grain and not along the grain.

On the other hand, Fig. 2.3 shows the failure of a 'plain' barrel made with a longitudinal seam so that the fibre of the metal runs along the barrel. Such barrels were cheaper to produce for military use but, as will be shown later, were theoretically stronger than Damascus barrels. In such a barrel the explosive forces within it, acting radially, stretch the tube fibres across the grain in their weakest direction, tending to separate them, rather than in the strongest direction along the grain as in the case of a 'twist' barrel.

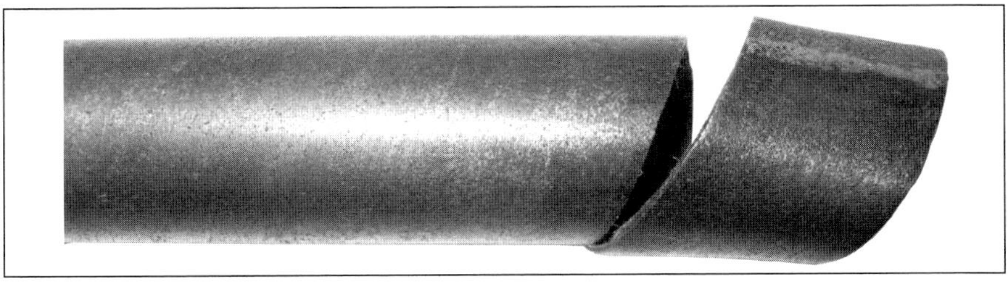

Fig. 2.2. A 'broken' twist barrel. The spiral pattern created by this method of manufacture can just be discerned and the break has followed the spiral path of the 'fibres' of the metal or the weld. (Author's collection © P. Smithurst)

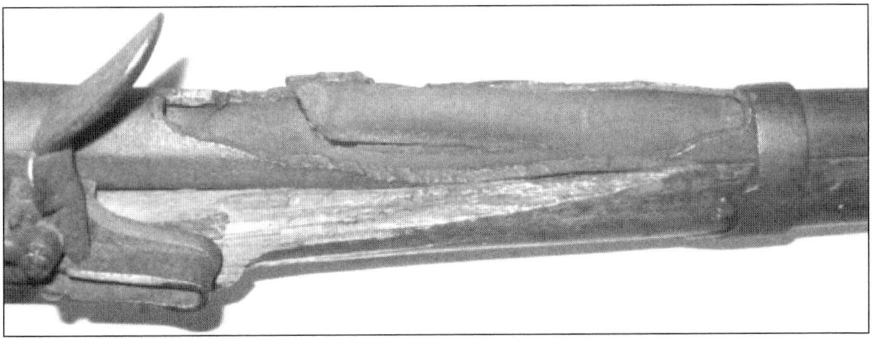

Fig. 2.3. A French musket with a burst barrel recovered after the battle at Waterloo, clearly showing the separation and fragmentation along the grain with degrees of tearing across the grain, possibly the result of double-loading by mistake in the heat of battle – a known mishap! (XII. 3141, © Royal Armouries)

Only later did the patterns of twist or Damascus barrels, revealed by polishing and etching with a variety of chemical agents, become valued, both for themselves and as a supposed mark of superior quality. These patterns are almost infinite in their possible variety because of the random microstructural nature of the materials from which they were made, ensuring that no two can be exactly alike. Later, as their popularity grew and gave them an air of prestige, they were loosely classified into different types. However, even these classifications, or at least the terms used to describe or define them, seem to vary between manufacturers but are not relevant to the musket barrel makers.

While these techniques for barrel making, with slight variations, were to continue well into the 19th century, there was a progression in manufacturing technology within this period, and for military gun barrels mechanised techniques began to evolve. This was to continue throughout the 19th century and is detailed in Volume 2 which is a study of the manufacture of the Enfield Pattern 1853 rifle. Where cost was not a factor, high-quality sporting gun barrels continued to be made using traditional hand-craft techniques well into the late 19th century.

2.2 Making forged and welded iron barrels in early 18th century Portugal

Espingarda Perfeyta contains a wealth of detail on barrel making in the early 18th century. It focuses on sporting gun barrels, and principally those with an octagonal breech, but is the earliest source of detailed information. However, it should be noted that whilst some military muskets as well as sporting guns might have had octagonal breeches, by 1707, in England at least, muskets were ordered to be fitted with plain round barrels[8] and became a feature common to most muskets of any nationality from this period onwards. Nevertheless, many of the techniques described are equally applicable to plain round barrels, both in the forging and finishing stages. A Portuguese barrel maker's workshop, in which two barrel makers are shown creating a tube, and before them are laid out the various tools used, is shown in Fig. 2.4.

Some of the terminology used is unfamiliar because of the separation of either time or geography, so it has been necessary to extrapolate to some extent in attempting to elucidate what is unfamiliar. According to the writers, three methods of barrel making were in use by that time, but what appears to be the oldest method was:

> *When our ancient Portuguese Masters wished to forge gun barrels, they took a bar of iron of the length that the barrel was to have, and proceeded to heat it and beat it equally, giving it the convenient width so that it exactly reached the length of the tapping iron* [taken to mean a mandrel] *on which it was to be forged: and then after being prepared, they made it all in the shape of a curved tile, and beginning in the middle of that bar on the tapping iron, they closed at one palm's distance, uniting top to top, or edge to edge, and having closed that part they put it in the fire, and continued to heat it as might be necessary for its security, and thus closing it so much more they repeated the same operation, and so by this method they took it to the point, and taking it by this after it was forged, they continued to the other point in the same manner, until it was completed.*[9]

The authors provide no illustration of this process so that in Fig. 2.5 is also inferred, showing the conversion of the 'tile', the channel shape into which the flat plate had been converted, into a tube with a welded seam.

Another method described in *Espingarda* for making long barrels was by the joining of several shorter tubes. The credit for this method in which the 'grain' of the iron lies circumferentially, is given to Juan Sanches de Miruena at the Court of Madrid. It is stated that he:

> *. . . used to draw the iron crossways, so that the iron of the barrel should have the grain across, for greater security, for, just as in wood or any board, there are to be found veins lengthwise, where it is*

Fig 2.4. The barrel maker's workshop (Fiosconi and Guserio, 1718, p.126)

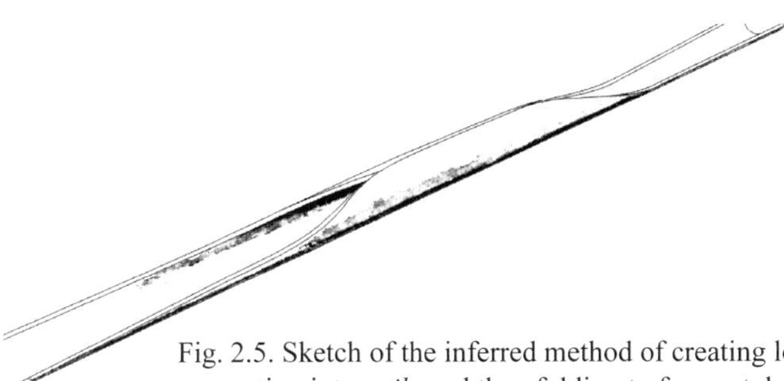

Fig. 2.5. Sketch of the inferred method of creating long tubes from a bar by converting into a *tile* and then folding to form a tube and welding the seam from the centre towards the ends, as described in *Espingarda Perfeyta*. (The mandrel or *tapping iron* has been omitted for clarity) (© P. Smithurst)

> *easy for cracks to occur, which does not happen across, whatever may be done, since it is against its direction. So, in iron in the same manner there are veins, which we call grain, and if these are across the barrel, it is stronger and safer. For, as the force made by the firing is equal in the whole span of the barrel, if the grain of the iron is lengthwise it easily opens, and for this effect any cause or origin is enough, but this will not happen if the grain is across, for even should there be cause and it be fired, it is no motive for bursting, for this does not weaken the barrel, and thus, as the fire does not find any weakness in length, it passes without damaging the barrel.*[10]

This passage affirms a total understanding of the fibrous nature of wrought iron. The barrel makers depicted in Fig. 2.4 appear to be engaged in making short tubes, Fig. 2.6.

A further detail from Fig. 2.4 shows the 'tiles' used in this process, Fig. 2.7.

The barrel-making process began with a wrought iron plate, often referred to in English terminology as a 'skelp', of the correct length, width and thickness, the cross section being in the form of a parallelogram. This plate was then converted by degrees firstly into a shallow channel, referred to as a tile on account of its shape, Fig. 2.7. It may have involved the use of the 'M' shaped device shown on the right of the illustration, Fig. 2.8, and which possibly functioned as a simple swage.

Fig. 2.6. Making short barrel tubes. (detail - Fiosconi and Guserio, 1718, p.126)

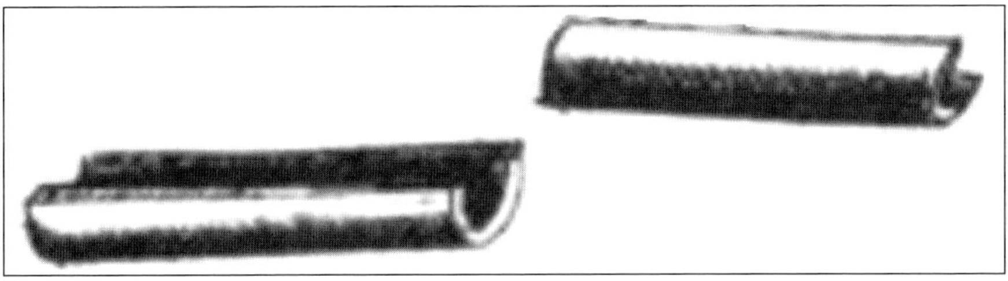

Fig. 2.7. Detail from Fig. 2.4 showing the 'tiles'. (Fiosconi and Guserio, 1718, p.126)

Fig. 2.8. Detail from Fig. 2.4 showing the 'M' shaped device and what appears to be either a round-faced hammer or a fullering tool possibly used to create the 'tile'. (Fiosconi and Guserio, 1718, p.126)

This would have been secured on the face of the anvil, the heated plate laid over it and hammered into the trough, either by striking with the round-edged hammer shown to the left in Fig. 2.8, or by using this as a 'fullering' tool by laying it on the plate over the cavity and striking it with a hammer. Then the upper sides were 'turned in' until the edges met and overlapped at the top, creating a long, scarfed, seam. This seam was then welded under the hammer with a round mandrel – the *tapping iron* referred to in the text – inserted into the tube to prevent the bore from collapsing under the hammer blows. Fig. 2.6 shows this latter process. If these tubes were only, say, 9 inches (225 mm) long, it may have required being brought to welding temperature two or three times. The whole process would have been similar to that shown in Fig. 2.9.

These short tubes were then joined by flaring the mouth at one end of the tube and bevelling the edge at the other to create a coned 'joint and socket' arrangement. Placing the bevelled end of one tube into the flared mouth of another, heating the joint areas to white 'welding' heat, inserting a mandrel to maintain shape, and welding the joint under the hammer would produce in effect a circular scarfed joint in exactly the same way as would be used in joining two lengths of bar together, Fig. 2.10.

Fig. 2.11 shows how they were configured so that the 'grain' of the iron was circumferential.

This is the earliest encountered acknowledgement of an understanding of the anisotropy in the physical properties of wrought iron, using it in such a way as to make best use of its strength with the 'grain' running circumferentially to better resist the explosive forces within the barrel. This property was first applied to artillery in the mid-19th century, utilising what became known as the theory of circumferential

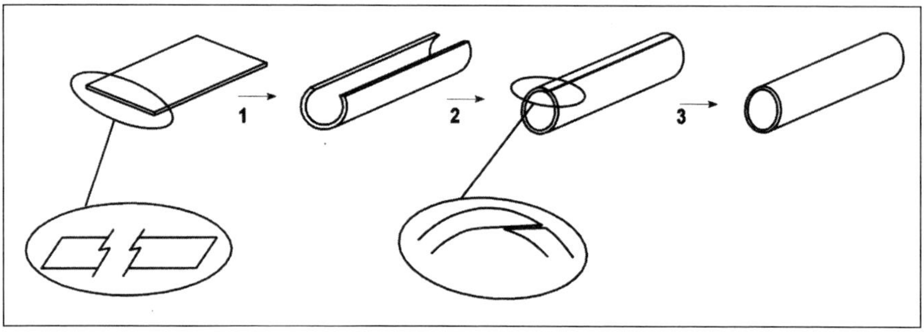

Fig. 2.9. Possible method of creating a short, welded iron tube: 1, forming a 'tile' from a plate; 2, closing up the tile to make a tube with an open seam; 3 welding the seam. (© P. Smithurst)

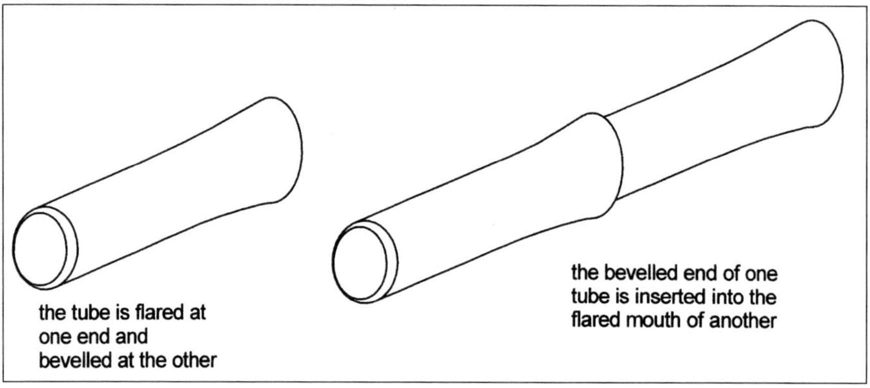

Fig. 2.10. Sketch showing the stages in making a long barrel from a number of short tubes as described in *Espingarda Perfeyta.* (The use of a mandrel – *tapping iron* - has been omitted for clarity). (© P. Smithurst)

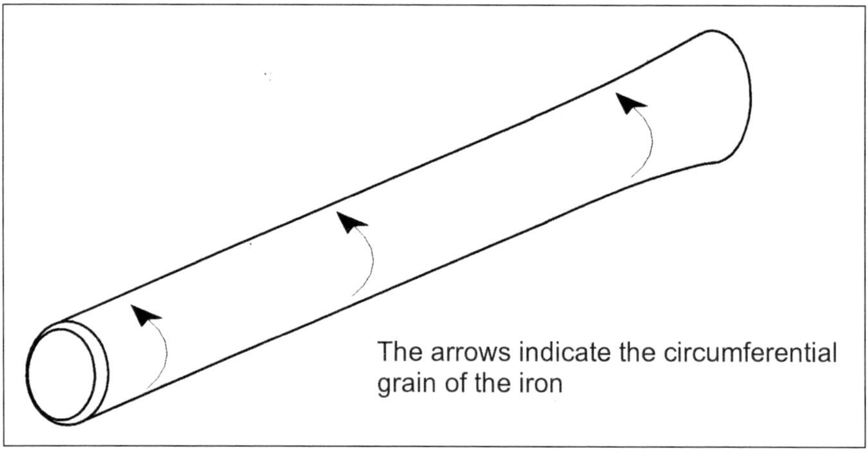

Fig. 2.11. Showing the configuration of the 'grain'. (© P. Smithurst)

tensions by Robert Mallet in the building of his 36-inch mortars as a means of improving the strength of artillery and making guns even larger and more powerful.[11] Surprisingly, Moxon, Fellow of the Royal Society and therefore a distinguished savant, in his *Mechanick Exercises* of 1703[12] makes no mention of this 'fibrous' or 'grain-like' property and its relationship to the strength of wrought iron.

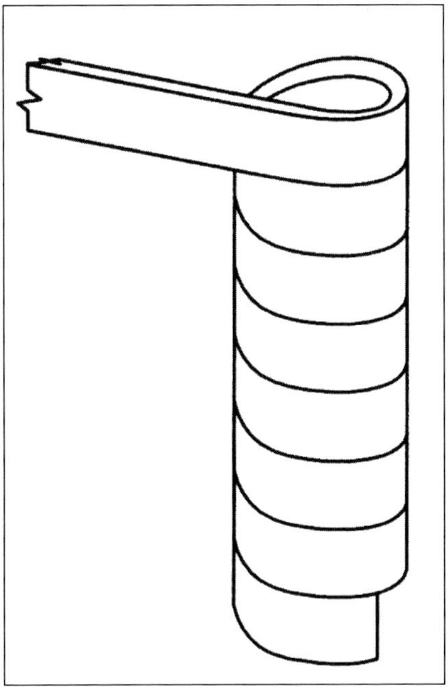

Fig. 2.12. Creating a coiled tube (© P. Smithurst)

This circumferential grain was also achieved in the third basic method described which produces what are referred to in *Espingarda Perfeyta* as 'wrapped' or 'twisted' barrels and which we are told was a procedure common amongst Catalan gunmakers at the time but was not ascribed to them for its origination.[13] In this process a strip or 'ribband' of iron, one inch wide and of such length so that when it was wrapped around a mandrel, the spiral coiled tube produced had a length of one 'couto', the width across the knuckles of the fingers and taken as approximating to 4 inches. What is clear is that the twisted barrel was also made in sections but, once again, no illustrations are provided so the illustration of the process in Fig. 2.12, while based on inference, has parallels in the English sporting gun trades through to the end of the 19th century at least.

Such barrels were regarded as being *extremely secure, as regards resisting the force of the gunpowder*, once again reflecting their level of understanding.[14] It was also observed that when bent, these barrels easily break. This is a further indication, and result of, the anisotropic properties of wrought iron. While the circumferential grain aids resisting the radial explosive forces by taking the tensile strain along the grain, bending places the stress across the grain where the strength is least. This may account for the damage to the barrel shown in Fig. 2.2 and once again, there is a direct comparison with wood.

The Portuguese gunmakers also noted that:

> *. . . after filing, all the rings of the strip may be seen around, and with these signs they are imperfect. That they may be secure, and resistant so that they do not break, and also that the joints may not be seen, another thinner strip is placed, after the first has been forged, on each of the additions, just at the beginning when the forging commences, joining one to another, in order with this second strip to close the first joints.*[15]

Fig. 2.13. Detail from Fig. 2.4 showing what appear to be gauges.

This description is unfortunately vague – is the whole barrel sheathed with this thin strip to cover all the spiral joints, or just at the points where the shorter tubes are welded together, and if so, would surely create another spiral joint? Such a question could only be fully answered by finding a barrel of this type and period and carrying out an X-ray or similar examination in the hope that the joints would be clearly revealed.

Mention is also made of making barrels using old horseshoes on the grounds that such iron is described as being very soft and workable.[16] This is similar to the later English practice in Birmingham of using old horseshoe nails, or 'stubs', which were regarded as having the same property and led to the term 'stub twist Damascus' for barrels made from these.

Throughout the barrel forging stage, simple gauges were used to ensure that the correct size and proportions were adhered to. What may be these gauges are visible at the bottom of Fig. 2.4, with rectangular and circular cut-outs, for the circular and octagonal portions of barrels, Fig. 2.13. Here again we are apt to think that gauges are a more recent concept but their use to ensure uniformity is such a simple expedient and since all craftsmen, in whatever field they worked, were adept at finding ways of making their task easier, we should not really be surprised at such innovation.

2.3 Making forged and welded iron barrels in the later 18th century - France and the M.1777 musket

The supposed universal nature of barrel forging is largely supported by accounts of practices in the later 18th century in France. Only three primary sources of information on barrel manufacture at this time have been found. Around 1777, Honoré Blanc, the French savant who is credited with attempting to re-introduce interchangeable manufacturing under Gribeauval in 1790, submitted his *Mémoire Important* to the French National Assembly which, amongst other things, noted in passing that he:

> *. . . devised a new way of forging gun-barrels, the result of which was to give the iron more density and render it more homogeneous by bending it more equally and more intimately. This new process gives the walls of the barrel quality which it is physically impossible to obtain by the methods which were formerly employed.*[17]

Unfortunately, as to the nature of this new method, no further reference was made by Blanc. Nor was it commented upon in the report issued by the Academie Royale des Sciences in 1791 assessing other aspects of Blanc's *Mémoire* and no further reference has been found. It is tempting to take his phrase - *and render it more homogeneous by bending it more equally and more intimately* - and view it as a process of repeatedly folding the iron back on itself and welding it in much the same manner as used in making Japanese sword blades. At each stage of this repeated welding more 'slag' would be ejected from the iron, increasing its density, and would also introduce a greater degree of, but never complete, homogeneity. In this it would be similar to the making of 'shear steel' in Sheffield by creating a stack of 'blister steel' bars and welding them together under a forge hammer. In the absence of further evidence, however, such an interpretation must remain conjectural.

Collectively, more informative are various editions of Diderot's *Encyclopedie* which, while providing some illustrations, offers little descriptive detail, and a comprehensive account by Cotty in 1806 but which lacks illustrations. However, while it was written later than Diderot, its descriptions match closely many of Diderot's illustrations, so it is possible to combine aspects of each. That fact alone indicates that in some respects, little had changed in the second half of the 18th century and, indeed, reflects some of the practices used in the early years of the 18th century. It should be noted that Cotty only refers to 'plain' musket barrels, as opposed to 'twist' barrels for sporting guns.

Cotty observes at the outset: *Before detailing the processes of the manufacture of the barrel, it will be observed that it takes two workers to forge it: the main one, called 'canonnier', brings the iron to welding temperature, which requires a lot of care and experience. The other operates the bellows and holds the mandrel ready to introduce it into the tube, at the moment that it comes out of the fire: both always strike quickly and in unison.*

It usually serves to have three tapered mandrels. They must be only 1.35 cm [0.53 inches] *in diameter at their thickest, so that if there are cavities or other defects in the bore, they can be removed by drilling without having to hammer the iron to force it inwards, which would be harmful. These mandrels, which serve to roughly form the bore are of a length relative to the distances for which they must be employed, and so as to always serve as support for the part which is to be welded.*[18]

This passage gives a clear acknowledgement of the problem faced by barrel makers in having to contend with small defects in the surface of the iron they had to work with, such as minute embedded cinders from the forge hearth, slag inclusions, blow holes, separation of the iron 'fibres' from being forged too cold and so on. Making the initial bore so much smaller allowed for these defects to be removed in the 'rough-boring' stage without having to resort later to hammering the barrel wall inwards at the point where the fault was located, creating a 'hump' in the bore to be removed by another boring.

The process then follows very much that already described for making a long tube in early 18th century Portugal: *the first operation of the barrel forger is to roll the skelp.* This is not what we envisage today by the term 'rolling' but, as Cotty notes, it was the conversion of a flat plate into a seamed tube, commencing with bringing the muzzle portion to a cherry-red heat and placing it *on an iron crutch,* [undoubtedly a mandrel] *adapted to the block that carries the anvil, where he makes it take a concave shape with hammer pein and then he finishes it closing it to the middle on the anvil; he then heats the breech end and closes it in the same way as for the muzzle.*[19]

This is interpreted schematically in Fig. 2.14 and is reminiscent of early Portuguese practice shown in Figs. 2.5 and 2.9.

These overlaps are shown more clearly in Fig. 2.15.

This method Cotty refers to as *forging by superposition* and as *being practised in almost all arms factories*. Because the barrel was reversed for working on each end, the seams overlapped from opposite sides, creating a 'dog-leg' in the junction of these folds. Two illustrations of a barrel tube formed in this way are shown in Diderot's *Encyclopédie*, Fig. 2.16.

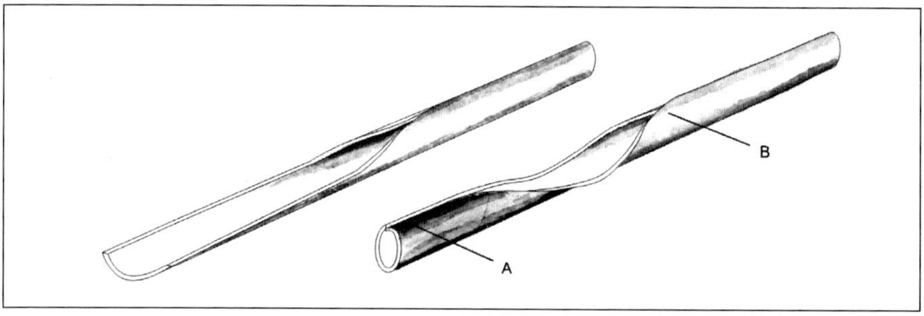

Fig. 2.14. Schematic representation of creating the tube in two stages by folding from opposite ends and with opposing overlaps, A and B, at the junction of the folds. (© P. Smithurst)

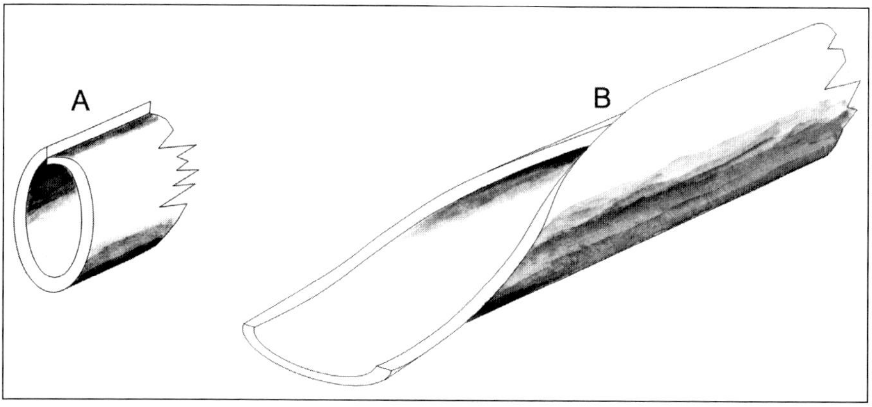

Fig. 2.15. The overlaps at A and B. (© P. Smithurst)

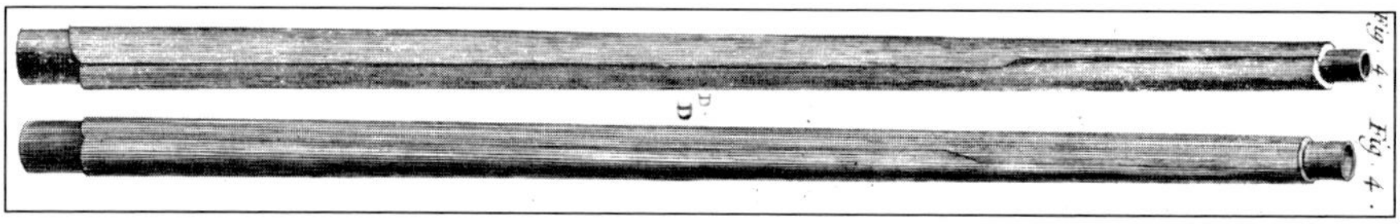

Fig. 2.16. Illustrations from two editions of Diderot showing the *lame roulée* (rolled skelp) with the point where the overlaps from opposite sides meet clearly visible as a 'dog-leg'. Top (Diderot, *Inter-livres*, pl.1 fig.4). Bottom (Pancoucke, 1784, pl.1, fig.4)

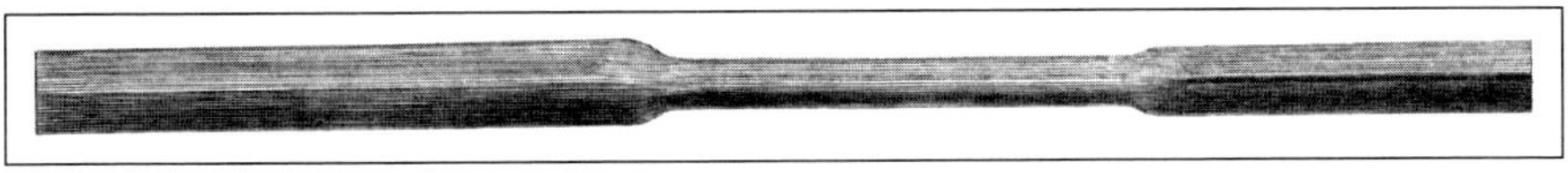

Fig. 2.17. Commencing to weld the tube, starting where the overlaps cross. (Diderot, 1784, pl. 1 fig. 5).

However, Cotty, in the same passage, also states that in their production: *The overlap is always . . . to the right of the forger when forging:* This is taken to mean 'facing to the right', in other words, overlapping from left to right and, since work commences at the muzzle, the top illustration in Fig. 2.16 is incorrect and is, in effect, a mirror image; there are other instances in this edition where images are 'mirrored'

Cotty then continues by noting that with the formation of the seamed tube completed, welding commenced where the two edges overlap. Again, this stage is illustrated in Diderot, Fig. 2.17.

After two more heats to weld the breech end in lengths of 5 to 6 cm [2–2.37 inches] at a time, the forger continued to weld to the muzzle. He was enjoined to take great care when bringing the central portion to a welding heat by supporting the outer end with the hammer when it was drawn from the fire because it was almost in fusion and could easily separate at this point.

The forger would also plug the end entering the hearth with clay or horse dung to avoid burning the iron and prevent anything entering the bore, and this was then removed to insert the mandrel. When heated and withdrawn from the hearth, the part to be welded would also be struck horizontally in the correct size of hollow cavity on the anvil, a process known as 'jogging' or 'jumping',[20] terms also used in the English gun trade, to bring the edges into closer proximity and partially unite them. This jumping was an important process in the production of 'twist' sporting gun barrels. When welding was completed, a small 'flare' was given to the opening of the bore at the breech and muzzle to ease entry of the drill. The final stage in the preparation of the tube was to heat it to almost welding heat and, without a mandrel, give it light blows from a hammer to *tighten the pores of the iron and to prevent defects*. At this point, after straightening by eye, it was ready for finishing.

2.4 Making forged and welded iron barrels in the early 19th century – Tula and the 1808 musket

The only detailed evidence from the early 19th century suggests that similar practices to those in France, and to some extent those used in Portugal in the early 18th century, were carried out in Russia at the Tula factory. Also, despite the lack of detailed accounts of gunmaking in English published sources, this Russian source, surprisingly, provides evidence that some of the processes detailed in *Espingarda Perfeyta* and those used at Tula for the manufacture of 'plain' barrel tubes, were also practiced in England. That a Russian report should contain such detail of English practice may seem strange, but it rests on the simple fact that in 1817 John

Jones, along with his son, John, from Birmingham were recruited to the factory at Tula and who, in the words of the author of the Tula report:

> *soon showed by his work his excellent knowledge concerning mechanics as applied to the manufacture of guns.*[21]

Jones junior was a die sinker and, although initially appointed to make dies for forging lock components, they also introduced improved barrel making techniques. Prior to Jones' appointment, barrels had been made at Tula in two parts and joined to create the full-length barrel, reflecting one of the techniques reported in use in Portugal. At Tula, two strips, 53.3 cm [21 inches] in length, Fig. 2.18, and weighing up to 6.3 kg [14 pounds], were placed in the hearth and brought to 'red heat'.

The muzzle portion was then elongated and tapered on the anvil. The second portion was elongated and tapered to a lesser degree; the thickest portion being used to form the breech. They were made so that the forward end of the breech portion was the same width as the rear end of the muzzle portion where they would be joined together. Skelps were sometimes referred to as 'boards' or 'plates', possibly derived from Germany where gunmakers referred to the skelp as a 'platine', whereas in French terminology, 'platine' is the lock or lockplate.[22]

Five smiths were needed for that operation, the master holding the metal in the tongs and four hammermen wielding hammers. One sometimes operated the bellows intermittently but occasionally an extra man would be used for that purpose. The forging of the two skelps took half an hour.

Converting each skelp into tubes was carried out in two stages on the anvil. The skelps were heated to bright red heat and, using a seven-pound hammer, the muzzle portion had its front half formed into a tube with the right side slightly overlapping the left side. It was then reversed on the anvil and the rear half was treated in the same way with the result that the tube had opposing overlaps as in the manufacture of the M.1777 barrels in France. The master held the tubes in the tongs and two men hammered them to shape. The manufacture of a breech tube and a muzzle tube took four hours.

Once shaped, welding of the tube seams commenced. A wooden stick was inserted into the muzzle end of the muzzle tube to act as a handle and the tube placed in the hearth so that the central portion was brought to red heat. It was then laid on the anvil and some 'squabbles' – the smallest chips from drilling the barrels – sprinkled on a length of several inches and the tube returned to the hearth and brought to welding heat. On being removed from the hearth, a 'crutch' – mandrel - with a 'T' shaped handle and having a shaft made of steel, was inserted

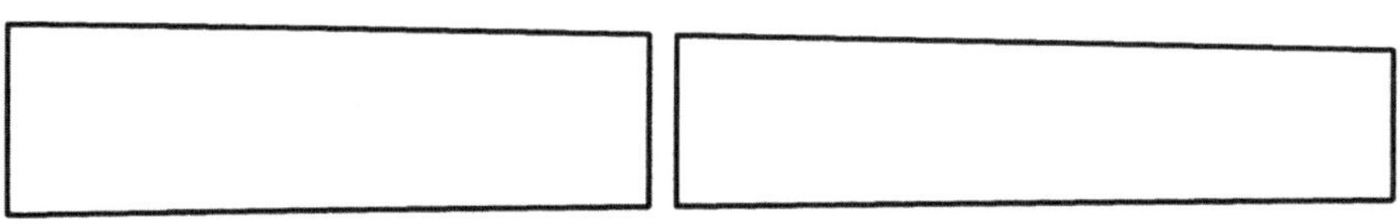

Fig. 2.18 – Sketch of the 2 skelps used to form breech (left) and muzzle portions (right) of barrels made at Tula. (© P. Smithurst)

into the tube and driven in tightly by several blows on a cast iron slab on the floor. The master laid the tube on the nose of the anvil and the hammer-smith struck it according to instructions of the master until the edges were completely welded. From time to time the crutch was withdrawn to prevent it becoming too tight in the bore. After heating this part once more, it was laid on the wetted anvil, a common practice to assist in removing any scale that had formed, and the welded edges given weak but very fast blows from the hammer to help consolidate the weld – again a reflection of French practice. This process was repeated, welding a few inches at a time. When one end was finished, the welded portion was filled with cow dung to prevent oxidation and the remainder welded in the same way. The whole process took about one and a quarter hours.[23]

After welding the muzzle portion, the two were united in the same way as described in *Espingarda Perfeyta*; the breech end of the muzzle tube was heated to a suitable temperature in the hearth and then flared on the beak of the anvil and the forward end of the breech tube fitted into it and then driven in by placing the breech end upright on the floor and giving the muzzle end several blows with a hammer. It was similar to the process depicted in Fig. 2.10.

The joint was then welded, followed by welding the seam of the breech tube in about twelve stages, taking around one hour. The whole process of forging the skelps and creating the welded barrel took about three and a quarter-hours, so a master could produce three barrels each day. Welding was carried out at night because it was difficult to judge when the barrel was at the correct welding heat, i.e., white-hot, in daylight and also difficult to observe any defects.[24] Anyone who has used colour as an indicator of temperature of iron or steel will know how difficult this is in bright light, whether natural or artificial. It also suggests that it was far better to weld barrels in the factory, as opposed to a forge at home, since in the former *it is possible to establish water-power operated cylinders for the supply of air* to the hearth to generate the necessary heat more quickly and surely.

In the translated version of Gamel's book, the opposed overlaps on the muzzle portion are reported as forming a 'cross' where they meet in the middle of the tube[25] but are more likely to have created a small 'dog-leg' in the welded seam. This seems identical to the detail on a barrel appearing in one of Diderot's illustrations as shown in Fig. 2.19. As in French practice, the breech portion was made with the overlap from one side only so only one of these 'dog-legs' would appear in the two-part barrel.

The arrival of John Jones and his son at Tula introduced a variety of innovations and improvements derived from English gunmaking practice, several relating to barrel making. One of the most important of these was the use of what are referred to as 'integrated skelps',[26] - a one-piece skelp of appropriate size and weight for

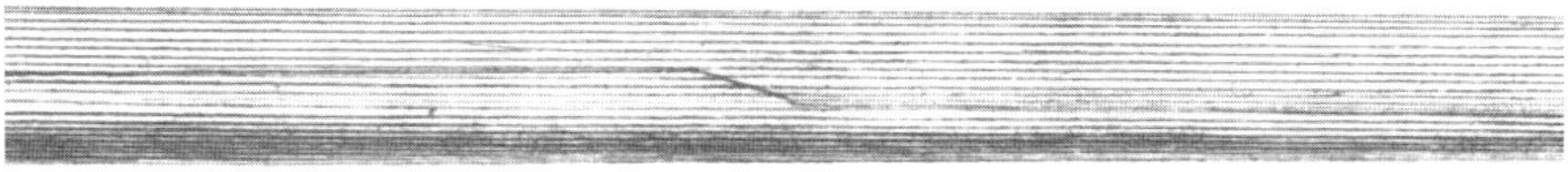

Fig. 2.19. Detail from Diderot showing the dog-leg in the welded joint at the junction of two opposing overlaps.

the making of a complete musket barrel. This probably reflects practice in England at that time.

Gamel states that Jones' 'integrated' skelps weighed 26.4 kg [12 pounds], were 107 cm [42 inches] in length, its width at one end 7.9 cm [3.1 inches] and at the other, 7.8 cm [3 inches]. Thickness at the wide end was 10 mm [0.4 inches] and at the other end, 3.8 mm [0.15 inches]. Russia adopted the English foot and inch measurements in 1831[27] so the arrival of the Jones' at Tula may have stimulated that change.

The wrought iron bars delivered to the Tula factory and from which the skelps were forged varied in dimensions. To ensure the correct amount was used, Jones used the simple expedient of a water displacement vessel which consisted of an upright cylinder closed at the bottom end and fitted with a tap close to the top edge. The tap was set in such a position that the volume between the tap and the brim equalled the volume, and therefore the correct weight, of iron needed. Water was added up to the level of the tap, so that when a bar of iron was immersed, when the water had risen to the brim, the bar could be marked with chalk at that point and then the surplus cut off.

To forge the skelps, Jones introduced a small water-powered hammer, weighing about seven pounds and traditionally referred to, in the Sheffield trades at least, as an 'oliver hammer'. The skelps were forged in three stages and therefore there were always two skelps in the fire whilst the third was being worked, and with the water-powered hammer three men could produce sixty skelps per day. The form of the integrated skelps, produced with bevelled edges, is shown in Fig. 2.20. Jones at this time was also developing three pairs of rolls for the production of skelps. One pair of rolls was for extending its length, one pair for increasing its width and the third and largest pair of rolls were cut with a tapered groove to give the skelp its proper shape.[28] However, rolling a relatively wide flat bar in respect to its thickness, and of tapering width to give it bevelled edges, (in effect creating a bar with a parallelogram cross section), without creating an undercut in the rolls would have been a difficult, if not impossible task.

Having prepared the skelps in the normal way, they had to be converted into tubes and this was done by using a purpose-made anvil and hammers also introduced by Jones. According to Gamel, the cast iron anvil had 10 semi-cylindrical channels in its face, Fig. 2.21; the largest was 10 cm [4 inches wide]; the smallest 3.3 cm [1.3 inches] wide.

The hammer used weighed 5.4 kg [12 pounds] and had one face of the head formed as a rounded ridge, in England usually known as a 'pein', set at 45^{0} to the

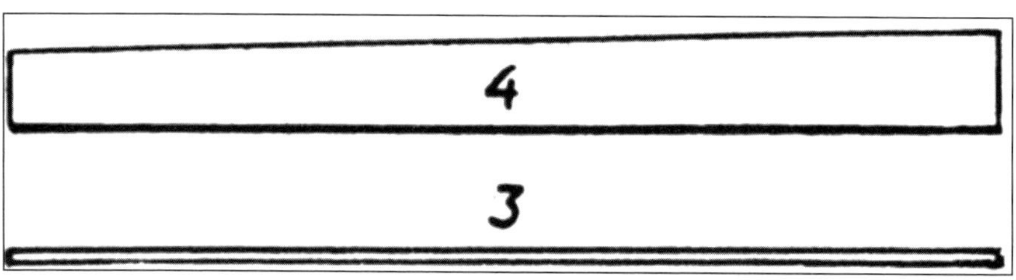

Fig 2.20. Jones' 'integrated' tapered and bevelled skelps. (Gamel, 1826, fig. IV)

hammer shaft. The skelp was heated to red heat in the furnace, laid in the widest channel in the anvil and using the angled pein of the hammer, Fig. 2.22, gradually formed into a channel. It was claimed that having the pein set at an angle ensured that the hammer would fall naturally with the pein more or less in line with the channel in the anvil, regardless of the stance of the user.

The Master had a similar hammer but weighing only two pounds and, although not stated, was most likely used to indicate where the hammer man was to strike with the large hammer. Having formed the channel, this, after heating and with the

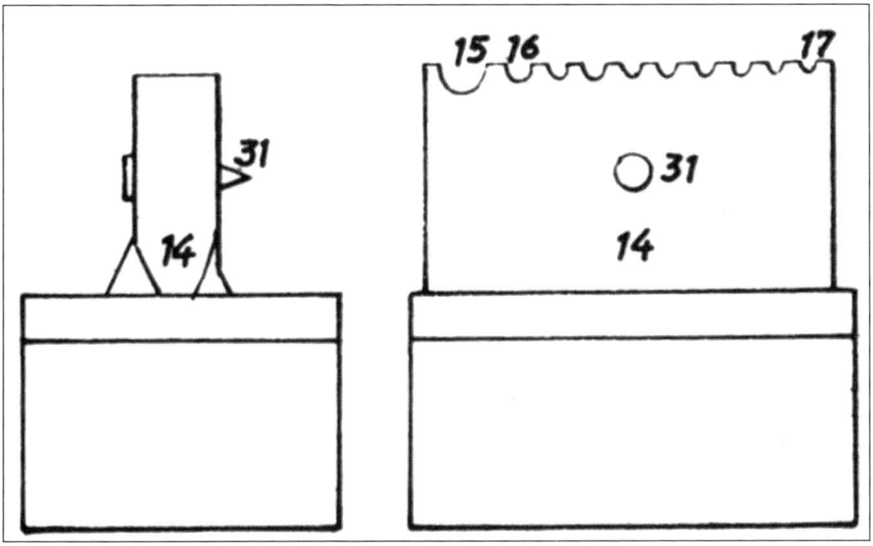

Fig. 2.21. Jones' *English style* barrel forging anvil at Tula. (from Gamel, 1826, fig. IV)

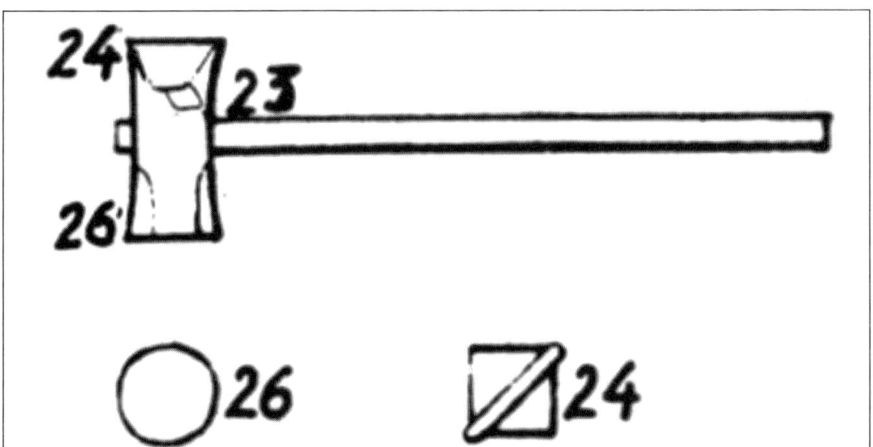

Fig. 2.22. The 12-pound hammer with a flat face, *26*, and a 'pein' angled at 45 degrees, *24*, used to form the skelp into a 'tile'. (from Gamel, 1826, fig. IV)

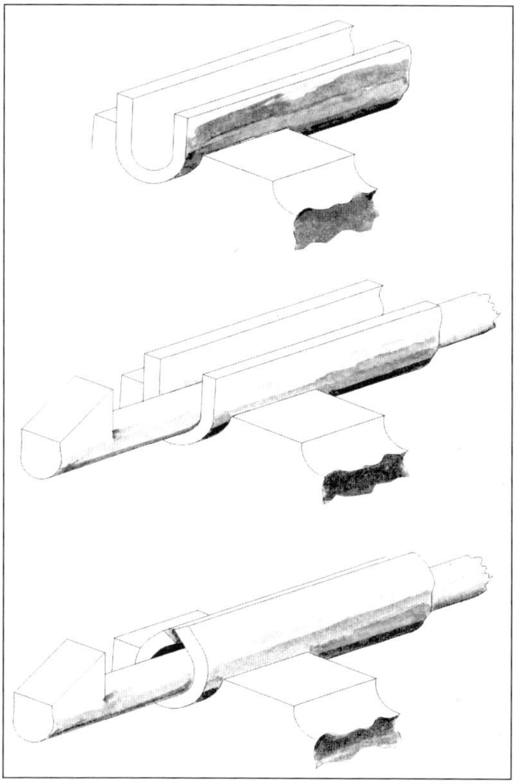

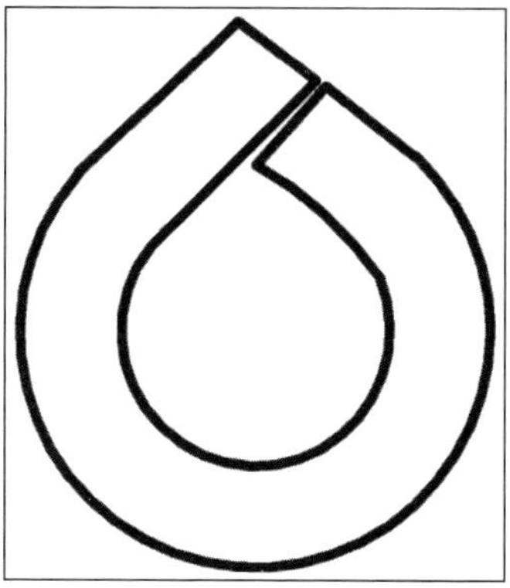

Above: Fig. 2.23. Cross section of the tube seam prior to welding.

Left: Fig. 2.24. Schematic representation of converting the skelp into a tube with the use of the large 'crutch' or mandrel. (© P. Smithurst).

large crutch, or mandrel, laid in it to maintain bore shape, was transformed into a tube using the flat face of the large hammer in Fig. 2.22. At odds with his earlier statement, Gamel states that the edges of the skelp were made square, not angled, so that when they were converted into a tube, one edge was butted against the opposing face as in Fig 2.23.

In a footnote,[29] Gamel claims that this was better than using a scarfed joint: *such thick edges of the tube are much better welded than thin ones (overlapping each other) made according to the previous method . . . cavities are formed in the barrels. These cavities are increased on drilling, causing a defect in the barrel.*

The whole process of forming into a tube is represented schematically in Fig. 2.24.

2.5 Making forged and welded iron barrels in 18th & early 19th century Britain

Considering that many techniques at Tula were introduced by the Jones' from Birmingham, it can be reasonably safely assumed, with regard to military barrel manufacture at least, that practices in Britain were comparable. However, numerous innovations can be found in the patent record. Two were of particular importance in regard to making 'plain' barrels for military use, those of Osborn and Heywood, and these are covered in the manufacture of the Enfield Pattern 1853 rifle.

Whilst some barrel making had been undertaken in London, by 1840 Birmingham held a monopoly, not only due to a concentration of the necessary skills but also because of the better coal available locally, having low levels of sulphur which was injurious to iron.[30] Various methods were used. It has already been noted that the military barrels were made from skelps folded into seamed tubes and welded. Even as late as 1840, and no doubt beyond, the method identical to the earlier method used at Tula before Jones introduced his integrated or 'full-length' skelp, was also used for sporting guns to be exported. W. Greener says of the barrel welders: *They proceed to weld probably a dozen long common barrels for the American trade, which are generally composed of inferior iron . . . rolled into two lengths of different thicknesses. These skelps are heated and beat on a groove until they form a tube half closed. They are then heated again, and closed with one edge over-lapping the other, as a brazier would over-lap the edge of a tin pipe for boys to blow peas with.*[31]

These seams of the front (muzzle) portion of the barrel were then welded in the traditional way. The end of the breech tube, where it would be joined to the muzzle portion, was flared slightly and placed in the hearth so that the parts of the breech and muzzle portions which were to be joined were brought to welding heat. They were placed on a mandrel and the joint welded in a few moments. The breech part then had its seam welded. However, if the breech part was to be octagonal, for instance, it would be welded on a flat surface, not in a circular groove.

There was little call for 'twist' or 'coiled' or 'Damascus' barrels, as they were variously referred to for military firearms, but they are found on Baker rifles, Fig. 2.25.

Fig. 2.25. Detail of a Baker rifle, ca. 1800 which, from the pattern of the barrel can probably be described as being a 'skelp' or 'single iron' Damascus to use Greener's terminology (Greener, 1884, p. 263). (Royal Armouries, XII.148)

However, despite their popularity for sporting guns and their perceived superior strength, Greener's comments on barrels made of any 'twisted' bars are not encouraging: *Iron is best when not twisted at all . . . for it resembles a piece of wood. The strands or fibres running parallel and firmly adhering, add strength to each other; if you twist them, you tear those fibres asunder, and they no longer support each other. So it is with iron. The objection made to the wire-twist is, that owing to the iron and steel being perfectly separate bodies, running through the whole thickness of the barrel, there is a difficulty in welding them perfectly, and, of course, there is a danger of its breaking across at any trifling imperfection, or mis-weld. This objection is certainly well grounded, as many barrels break at proving. I have myself seen a very strong barrel indeed broken across the knee, without the slightest difficulty, while, to all appearances, it was perfectly sound. This is the reason why the manufacturers have ceased to make them, except for the American trade. . . . In short, I hold a Damascus to be little superior to a common barrel, where the fibres run parallel to the bore.*

He goes on to give some comparative tensile strengths but the figures, or more correctly the terminology he uses, is rather confusing today:

> *. . . experiment with the testing machine, shows a rod of wire twist* [this is taken to mean the wire used to make the twist] *3/8 inch square, containing 1.6875 solid inches as equal to 11,200 lbs. A rod, when converted into Damascus of 11/16 of an inch in breadth, by 4/16 inch in thickness, containing 2.625 solid inches, was only equal to 8,960 lbs, being thus a clear loss of full 35 per cent.*[32]

What is meant by 'solid inches' is the volume of metal in the test piece, but that has no bearing on the tensile strength. A more meaningful approach is to take these figures and convert into the more usual way of expressing tensile strength as weight per unit cross-sectional area.

In the first instance, this computes as 35 tons per square inch; in the twisted rod it becomes 23 tons per square inch, giving the reduction of 35% Greener claimed. He gives similar figures for 'three-penny skelp' and 'two-penny, or Wednesbury skelp' which compute as 27.1 and 17.6 tons per square inch respectively, a serious reduction in strength considering their intended purpose. He does not even bother to give figures for 'sham damn skelp'!

2.6 Making forged and welded iron barrels in 18th & early 19th century America

In America, at Harper's Ferry and Springfield Armories, the same basic techniques of military barrel making were being used.[33] There were, after all, only a limited number of methods that could be used.

Chapter 3

Finishing the Barrel Internally

3.1 Introduction

. . . well proportioned, and that the thickness and length of the breech (whether it be octagonal or round) accord with the rest of the barrel, so that seen altogether it has equal and well drawn proportions. Besides these circumstances the most important of all is that the barrel be well evened so that it is not in one place thick and in another thin, for this equality and perfection concerns the safety of the barrel itself, and the others satisfy only appearance.[1]

To this might be added an equally valid comment:

> *. . . the degree of straightness required in a gun barrel is nothing short of absolute perfection.*[2]

The opening statement by the authors of *Espingarda Perfeyta* reflects what must have become apparent very early in the gun's history and is equally true today; that no matter what type of gun, it was perfection in the physical and mechanical attributes of the barrel which was paramount, regardless of aesthetics. Simple reflection will show, for instance, that if the barrel wall is thinner on one side, then that could constitute a serious weakness and the barrel may burst. From the more practical aspect, if they are not straight *and* coaxial, any sighting arrangements for aiming and hitting the target would rely more on guesswork and good luck than on skill and good judgement. The technology by which these objectives would be ultimately achieved was through the evolution and development of a variety of machines with a dedicated purpose.

Thus, the making of the 'tube' was only the first stage in the process of producing a barrel. It then had to be finished, inside and out, using machines whose mechanical sophistication gradually advanced, but work on the bore and the outside of the barrel often alternated so there is no convenient way of separating 'internal' and 'external' operations. The subject will therefore be treated in a more or less chronological sequence.

3.2 Finishing the bore in early 18th century Portugal

As the writers of Espingarda point out, when finishing the barrel, the first process was to create a bore which was perfectly smooth, uniform and straight. This is in keeping with much of current engineering practice where, if an accurate bore is required, it is created first and acts as the datum for all further operations and measurements.

Another different view of the part of the workshop where barrel drilling/boring operations were carried out is shown in Fig. 3.2. The writers also fully appreciated what was needed in a machine to accomplish this - *The lathe shall be made with every precision and perfection, and its wheel shall be larger, rather than smaller, firmly set with the circumference so exact, that in whatever part it may be desired to stop it, it remains motionless. The chuck, where there enters the handles* [shanks] *of the drills, must be very straight, so that these also work straight without inclining to any side, and it shall also have its hole, wherein the catch is put so that the drills are not released . . The bench [bed] . . shall be 15 palms* [60 inches, 1524 cm] *in length, and one in thickness* [4 inches, 10 cm], *and it shall be very strong, with the four legs buried deep, so that it is quite firm, and does not shake or move. . . . Good*

Fig. 3.1. Part of a barrel maker's workshop showing, lower right, the horizontal drilling lathe. (Fiosconi and Guserio, 1718, pps. 8, 9)

care shall be given to the setting of the bench, and the wheel, which must be very straight and exactly at the height of the wheel shaft . . .[3]

Maudslay could not have put it better and, despite a structural reliance on wood, they reflect some of the principal attributes of the lathe today. It also removes any doubt that the items shown hanging on the rack at the rear of the workshop illustrated in Figs. 3.1 and 3.2 with holes at their extremities are drills and not mandrels. (See Figs. 3.6 and 3.7 for details)

Fig. 3.2. The barrel drilling workshop. (Fiosconi and Guserio, 1718, p. 76)

The horizontal boring machine, Fig. 3.2, was not an innovation. Whilst an earlier illustration of a machine for use on smallarms, i.e., hand-held firearms, has not so far been encountered, there are earlier illustrations of those used for cannon boring (Fig. 3.3).

Obviously, it is larger than the drilling lathe shown in Fig. 3.2 and is powered by either a waterwheel or an animal-driven treadmill, but closer comparison with the aid of some details will show that the same basic principles are embodied in each. Biringoccio's illustration is largely self-explanatory but, as depicted in Fig. 3.4, the cannon barrel, *B*, was secured on a carriage, *C*, which could be made to slide in either direction in a controlled manner along a substantial wooden bed by means of ropes, *R*, attached to it front and rear and to windlasses *W*, (only one shown). By that means, the cannon could be fed onto, or away from, the rotating boring bar or drill, *D*, fitted to and driven by the large driving wheel on the left. The contrivance at the upper right with what appears to be a large spoked handwheel is simply to show the way in which the 'driving wheel' with its boring bar/drill bit is supported.

This same operational system is shown in the boring machines featured in *Espingarda Perfeyta* but variations in terminology and language over the three hundred years since it was written have led to curious terminology in translation. These, combined with some lack of clarity in the illustrations, mean that some aspects of the description of the machine and its use are confusing. The description here therefore, along with the schematic prepared, Fig. 3.5, is a synthesis of what can be gleaned from the description and deduced from the illustrations.

Thus, in the barrel drilling machine as depicted in Figs. 3.4 & 3.5, the barrel tube, *T*, is mounted on a carriage, *C*, which can be made to slide along the bed of the machine by means of ropes, *R*, attached to each end. These ropes, instead of being attached to some form of windlass, each pass over a pulley at the end of the machine bed, then over a pulley attached to the outer end of a pivoted board, *B*, and

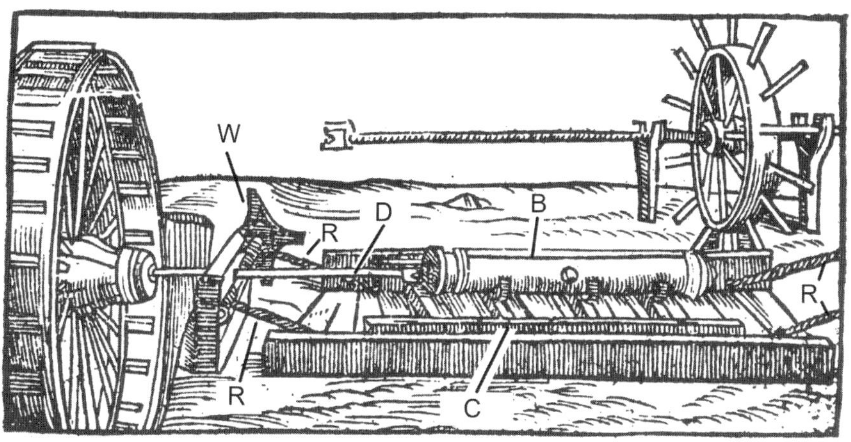

Fig. 3.3. Cannon boring machine of 1550's with the addition of key letters. (Biringoccio, 1558, p. 113)

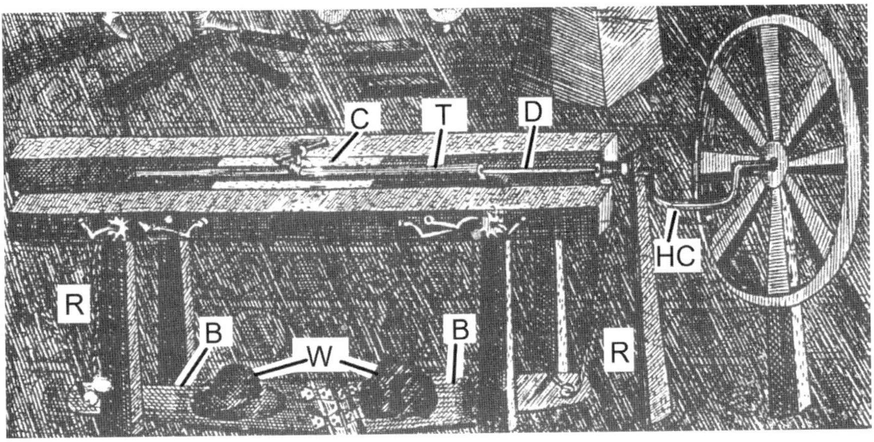

Fig. 3.4. Detail of barrel drilling lathe shown in Fig. 3.2, adapted by the addition of 'key' letters.

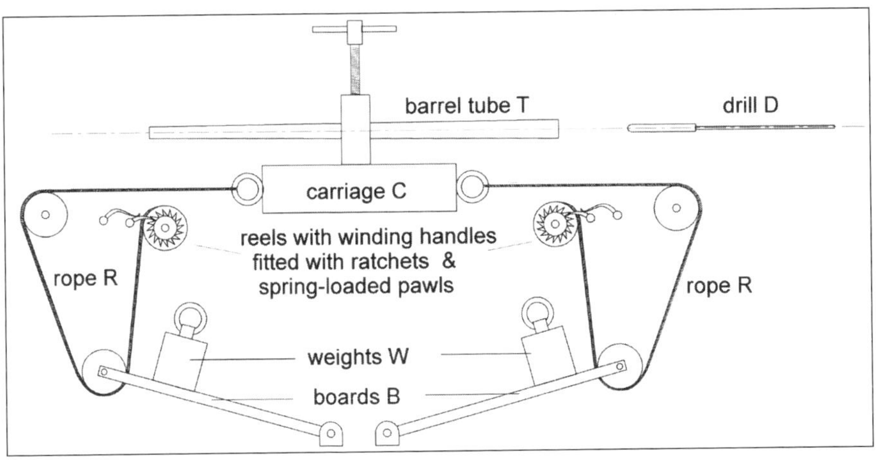

Fig. 3.5. Sketch showing the mechanical elements of the *Espingarda* barrel drilling lathe. (© P. Smithurst)

finally to a reel or winding drum operated by a handle and controlled by a ratchet wheel and pawl. Weights, *W*, sit on the hinged boards, *B*, which exert a tension on the ropes, *R*, so that, by releasing one or other of the reels by disengaging the ratchet and pawl mechanism, the carriage is drawn in one direction or the other while the drilling bit, *D*, is working inside the barrel. Unlike the case of boring a cannon barrel, the drilling bit in this instance is powered by being attached to the shaft of a flywheel and operated by the hand-crank, *HC*, Fig. 3.4.

An interesting aspect of this machine is that by sliding the weights along the boards, *B*, the tension on the ropes can be varied and, thereby, the rate at which the drill is fed into the barrel. It does not stretch credibility to imagine that whether firearms or cannons had first inspired the creation of the boring machine, makers of the other would adapt it to their needs, as indeed the cannon boring machine was later adapted to bore steam engine cylinders.

In connection with the drilling/boring operation we are told there are - *twelve drills with square cutting edges made of good steel, each diminishing in thickness by one adarme, to which end measurement shall be made by the calibrator, and their shafts little thinner than their cutting edges.*[4] [An 'adarme' is a Portuguese measure of weight, in the same manner as bore size being equal to the number of lead balls weighing 1 pound, the adarme equates to .08 inches [2 mm] approximately.]

The use of drills with square section cutting heads would seem to be associated with a variety of factors. The principal advantage, and a very important one, is that the cutting head, being symmetrical and relatively long, would, to some degree, have been self-centering. In addition, it would have been quite stiff and, by using a succession of incrementally larger drills which removed only small amounts of metal, would have tended to follow a straight path, removing slight irregularities in the process of passing through the bore.

On the other hand, the square section meant that the corners, which formed the cutting edges, were attempting to cut at a far from ideal very obtuse angle. This is more in the nature of a reamer and is in stark contrast with a typical cutting tool used in a lathe, for instance, which has an acute angle at the tip to create a sharp 'point' and whose top face is at an angle of about 10 degrees – the 'rake' – downwards from the centre line of the work. This enables it cut into the surface of the metal and 'peel' it away, compared with the square drill which has more of a scraping action, Fig. 3.6.

It could be thought that a better cutting action could have been obtained from an equilateral triangular-section head on the drill which would have at least

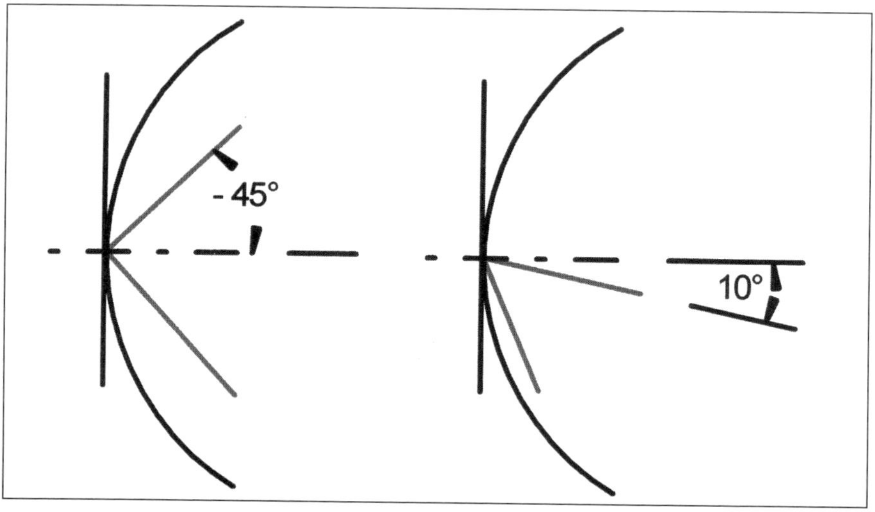

Fig. 3.6. Shown in grey - on the left, the obtuse angle of contact of the square drill compared with the more usual acute contact angle, right, of a typical cutting tool used in a lathe for boring or turning. (© P. Smithurst)

reduced the angle of contact at the cutting edge, but it would still have acted with a 'scraping' action.

The 'calibrator' mentioned above appears to have been a form of master gauge relating to the bore sizes and against which plug gauges are made and checked:

> *He shall have a calibrator with the correct calibres of the weight of each of the bullets, and it shall have twenty holes, well set apart, and for these there shall be twenty iron gauges, each of them with a point two fingers long, and on those points shall be a thread, for screwing into a nut, which shall be fixed into a small wooden rod.*

There are two devices illustrated, Fig. 3.7, which approximate to the description of the *calibrator*. They differ, however, in terms of the number of holes; as can be seen, the one on the left has 18 and the one on the right, 8, against the prescribed 20 and may be simply 'representative'.

The description continues: *And to make this calibrator, twenty lead bullets shall be obtained, well rounded, and the largest shall weigh twenty adarmes* [2.5 ounces, or 70g., so between 6 & 7 bore], *another that weighs nineteen, another that weighs eighteen, and in this way he shall have bullets that diminish by one adarme until the twenty bullets fit exactly into the weight, as we have said. These bullets are made in well-rounded shapes, and as on their roundness depends every good circumstance, it is better that they be made on a lathe.*

After these bullets are made the holes must be enlarged in the calibrator with a six-edge bit [this sounds similar to the clockmaker's broach which, in English practice, was slightly tapered and usually had five sides, and functioned as a reamer], *causing the first bullet, which must be greater, to enter exactly, and thence onward to the last, opening exactly on a decreasing measure, while at the side of each hole, by its numbers, is marked the figure of each of its calibres. Then into this, and its holes, shall be fitted the iron plugs. It is enough for these to be round in the middle circumference, and being worked on the lathe they will be more perfect and thus fit well into the calibrator.*

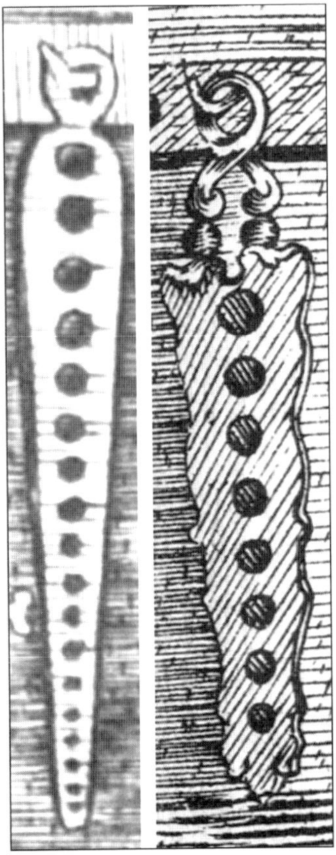

Fig. 3.7. Details from Figs. 3.1, and 3.2, showing what may be 'calibrators'.

These plug gauges were then used so that, *when the gauge is within the barrel, by putting it up to the light it will be seen if the barrel is well rounded. These gauges must be numbered in a similar way to the calibrator.*[5]

The gauges were therefore used not only to confirm the dimensional accuracy of the bore, but they were also being used to examine it for 'circularity' and 'cylindricity'. As already noted in the previous chapter, gauges were also used in the making of the welded tube for verifying or controlling sizes and together these are the earliest documented use of gauges in gunmaking so far encountered.

The drilling was started with four of these drills in succession, the heads of which were one *hand* in length, approximating to 4 inches (see Appendix 1), tapering to the point and with a shaft long enough to allow them to pass right through the barrel.

The next four were to have, *the length of two more fingers* [approximately 1½ inches / 45 mm] *on the cutting edges, and be less tapering to the points, and the length of the shaft sufficient for it to come out of the barrel.*

The last four drills were to have, *the same length, with the difference, however, that the edges shall be one palm long, and even less tapering to the points, because they have to be used to perfect the barrel, and so must have their heads well finished and almost equal, and to this end they shall be filed according to the set square, rule and cavedaes,*[6] *and after being tempered and sharpened, shall be quite equal and exact as before, without roughness or too thin an edge, or softness, for with any of these defects they will not do good work.*

Finally:

The last drill is that which serves to polish the barrel [bore], *and for this purpose must have its edges sharper, so that may serve for polishing, and they must be tempered to a blue colour. Care shall be taken that the sharpening of these drills be done with well finished stone, and so that they shall have the edge fine, it will be better to use stone that is white, for this is by nature harder and polishes better. With the rule the edges shall be tested to see if or not they have high and low points, so that these may be removed, and the edges be made quite uniform.*[7]

There are thus thirteen drills used in the process described. However, the illustrations show different numbers. Examining Fig. 3.1, for instance, only nine items which correspond to drills are hanging on the left of the rack.

On the other hand, Fig. 3.2 shows 15 drills!

Also, in neither illustration are they all shown to be of a length that would 'pass through' a long barrel such as might be found on a sporting gun. These apparent discrepancies could all be ascribed to 'artistic licence' but that seems out of place when the authors have been at pains to provide accurate visual detail elsewhere.

What we might see as a lack of clarity, as commented upon elsewhere, is possibly due to translation three centuries removed from the time of writing, in which terminology or language may have changed.

Unfortunately, the obscurity does not end there. We are next told:

> *For these drills, which pass through the barrel from one end to the other, there shall be rods*[8] [reeds] *which are made of walnut wood to*

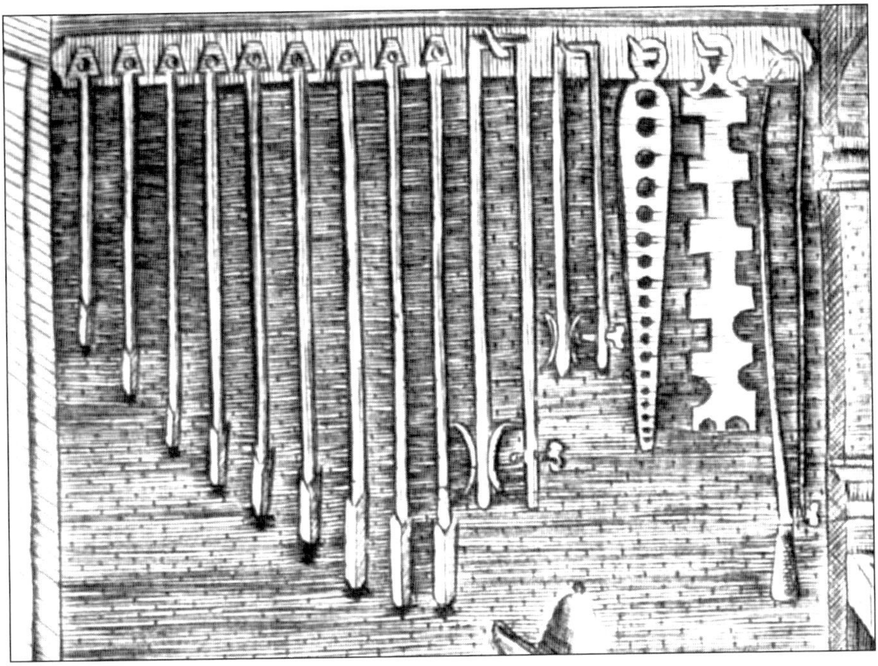

Fig. 3.8. Detail from Fig. 3.1 showing only 9 of what are taken to be drills hanging on the rack.

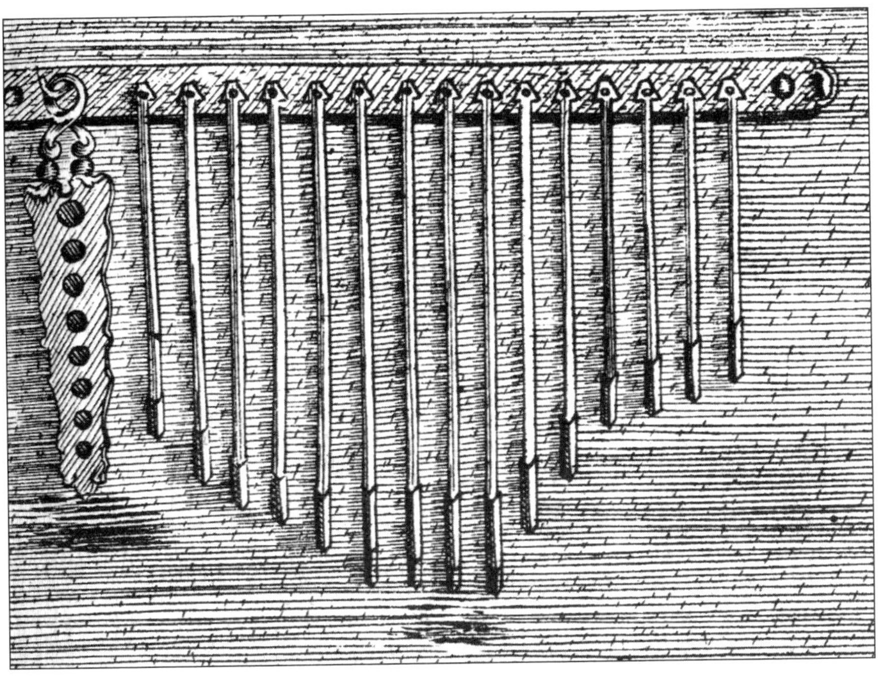

Fig. 3.9. Detail from Fig. 3.2 showing the fifteen drills.

> *cut, and others of chestnut to polish, of the length of the same cutting edge, hollow inside, which receive in the middle an edge of the drill, the two sides being covered, so that only one edge is used for cutting, and these rods shall be covered at the points.*[9]

Despite having corrected a possible error in the translation, it is difficult to comprehend what this passage is trying to convey. It seems to imply a drill with 3 edges, not four. It is not helped by that fact that these processes as described seem at variance with a later section of the book in which the process of drilling and straightening the barrels is described. It may, of course, simply be that this later section, Chapter XXI, elaborates upon what was previously an outline or general overview of the process.

Chapter XXI commences with the statement: *In order to drill the barrel within, it must be very straight, and firmly held in its place in the lathe, and beginning to drill at the breech with the short drills, the first being the thinnest, work shall continue with the others which follow it.*

It then continues: *the drills shall be put into it* [the barrel] *in the manner described, and they shall be worked until seven adarmes are reached, everywhere even, and then long drills shall be used.*[10] [author's underlining]

It will be recalled that in the earlier section describing drilling, only long drills, i.e., those capable of *passing through* the barrel were mentioned.

Moving on: *On these shall be put their rods* [reeds], *closed in such a way that the drills cut with only one edge, and they shall be passed through the barrel from end to end, until it is even throughout, and clear, so that thus straightening with the cord may be begun.*

Despite the confusion regarding which drills were used and how, the checking and correcting for straightness is very clear. The *cord* which was used to check the straightness of the bore, Fig. 3.10, was a very thin viola string with weights at each end and, so as not to damage it or cause any error, the edges at the extremities of the bore were filed smooth. The barrel was then rested on two supports. Since the breech-end was octagonal, it was rested on one of its flats which acted as a reference surface. The barrel was pointed towards a window so that the interior was illuminated for viewing:

> *Thus, it is possible to see the hollows in the barrel, and if there are any such, they are put right with the hammer, striking on the outside, and when they are large, the barrel shall be bent by hand in the groove which for this effect shall be in a wooden bench.*[11]

The 'viewer' is holding in his right hand the *probe* which is simply: *a wire rod three palms in length,* [this is clearly an error since the illustration shows a probe that is more appropriate to a long barrel] *not very thick, which has one end turned down, and this probe shall be used to measure within where there is any hollow, or gap, on which the cord does not lie, and then place it outside, and at the corresponding*

Fig. 3.10. Detail from Fig. 3.2 showing the barrel being checked for straightness with the cord.

place a mark shall be made with the file where there have to be given the blows to straighten it. This probe also serves to find the place within where there is any prominence, to show outside the point where there is need to reduce it with the drill.[12]

The barrel was checked in this way, laying it on all eight faces of the octagonal breech and viewing from both breech and muzzle end. Any defects discovered were corrected either by laying the barrel on a wooden bench and striking a judicious blow with a hammer, or, if the discrepancy was large, by placing in a groove in the bench and bending judiciously by hand.

The barrel was then once again placed in the drilling lathe and, using a long drill, *arranged so that it cuts with only one edge, and thus the prominences made by the hammer shall be smoothed, always using oil from beginning to end, together with a wet cloth always on the barrel where the drill is working, so that it shall not heat and be scored by the drill, which easily happens when the barrel is hot. If it be desired that the rod* [drill] *fit more tightly, there shall be put a strip of folded paper between it and the edge of the drill and this work shall continue.*[13]

What is presumed to be meant here is that the paper shall be put opposite the cutting edge of the drill, causing the drill to cut a fractionally larger diameter –

a technique certainly continuing through the nineteenth century in English gunmaking practice.

This drilling, followed by checking *by the cord* was repeated until the correct calibre had been reached and the bore was perfectly straight and round.

Once this point had been reached, the final drill was used:

> *that has the edges sharpest, and so well sharpened shall the edges be, using the white stone, that when passing them on the fingernail they scrape it.*

Additional instructions were that: *care be taken that on the cutting edges there be no bristle* [burr] *or any other impediment, for if there be they will not do good work, and thus, before putting the drill on the rod* [reed], *the edge shall be run along the nail. And this step shall be taken continually on putting it on the rod* [reed], *so that the barrel become bright within, and have no defect.*

From the continuation of this description, it would appear that, after this process had been carried out satisfactorily, and prior to the bore being polished, the barrel was:

> *tried with three charges of gunpowder and three bullets, in order to know if inside there is raised and little scale. If there be any, the probe shall be used to probe where it is, and at the mark that is made outside, there shall be given some small blow, sufficient for it to be smoothed by the drill, and when this has been overcome and is adjusted, it shall be prepared with the polish.*

It is not clear whether this means three separate firings or one using a triple charge and 3 balls fired together, though this could be excessive. [After finishing the outside, a double charge and two bullets were used to 'prove' the gun.][14]

The description continues that for the polishing of the bore:

> *On the same drill, or any other which passes through the barrel from one end to the other, there shall be put enough rods* [reeds] *to cover the cutting edges, so that it enters the barrel somewhat tightly with the rods* [reeds], *which shall have on them emery, well washed so that it is finer, and with oil, they shall run up and down the barrel so many times as are enough to make it bright.*

Once the bore was polished, the breech plug was hardened and tempered and its face exposed in the bore polished with Milan polish, a very fine grade of emery powder, so that when the bore was viewed from the muzzle, it could be seen even when the vent was blocked.[15] The work on the bore was finished and there is no suggestion that any external machining was conducted prior to this point.

3.3 Finishing the bore in the later 18th century - France and the M.1777 musket

However confusing the *Espingarda Perfeyta* narrative might be, the clarity in some details indicate machines were used that were similar to those described later by Diderot and those used at Tula, which had been imported from England in the early 19th century, and that were still in use in England into the later 19th century for the production of sporting gun barrels. What is remarkable is how little these had changed in their basic operation in the space of one hundred and sixty years and a study of these can provide a retrospective clarity to *Espingarda Perfeyta.*

Illustrations by Diderot in 1762, Fig. 3.11, and 1779, Fig. 3.12, exhibit differences of scale from *Espingarda* and in the source of power for the machines.

They embody basically the same technology as later described by Cotty in 1806:

> *The barrel is drilled in a workshop which has a number of lantern gears* [headstocks] *proportionate to the power of the water, to the diameter of the driving wheel, etc. These lantern gears, placed horizontally and parallel to each other, each carry a drill; they mesh with a number of vertical* ['crown'] *gearwheels carried on a single shaft. At the end of this shaft, is a large horizontal lantern gear that receives the movement of a tall vertical pin gear, carried by the same shaft of the water wheel. The drill is exactly fixed in the center of the lantern gear which gives it its horizontal rotation, and it passes in the barrel which is placed on the same axis: the barrel is fixed on the drill bench where it is advanced onto the drill in a uniform way.*[16]

The 'lantern gears', so called because their construction with circular plates connected by bars is reminiscent of a lantern, acted as a simple headstock into which a drill was 'plugged', Fig. 3.13.

In Cotty's words: *The drills, which it is better to call reamers, as in other arts are used, are of steel, tempered, the shape of a truncated quadrangular pyramid, cut with the four edges, and welded to a rod of iron, which is flattened at its end*

Fig. 3.11. Barrel drilling and boring machines. (Diderot, 1762, pl. 1)

Above: Fig. 3.12. Barrel drilling and boring machines. (Diderot, 1779, pl. 3)

Left: Fig. 3.13. A 'lantern' gear. (detail from Fig. 3.12.)

which enters the socket in the centre of the lantern gear: the smallest of which is 1.13 cm at its greatest thickness, and the largest 1.67 cm thickness.[17] [The full measurement notation was used by Cotty since at the time he was writing, 1806, it was a period when the metric system was being introduced alongside the old 'imperial' measure, but has been omitted here.]

From this description of the drills, and images of the machines, they bear close resemblance in their basic technology to those used in Portugal, Tula and in England throughout this period and later, except that Cotty's drills cut on all four edges, as opposed to a single edge.

Often, it was found that the barrels warped during drilling due to the heat generated by friction and the workman had to regularly pour water on it, as well as applying oil occasionally. The details of the machine shown in Diderot's illustration of 1762, Fig. 3.11, are provided in Fig. 3.14 which shows the drill, *Q*, fitted into a socket, *P*, on the driving spindle (*lantern gear*) and operating in the barrel. The barrel itself is fitted to a recess in a sliding block, *T*, so that its centre height coincides with that of the drill, and is held rigidly in place by a clamp, *S*. Cooling water is fed onto the barrel during drilling by the conduit, *C*. Item *t*, though no mention can be found, is presumed to be part of the collection of twenty-two graduated drills mentioned by Cotty.

In Plate 2 from Diderot's 1779 edition, Fig. 3.15, translations have not been found for names of some items relating to the key letters; thus, the barrel, *Q*, is fixed to a *sepay* (taken to be a carriage?), *O*, which slides along and is guided by the

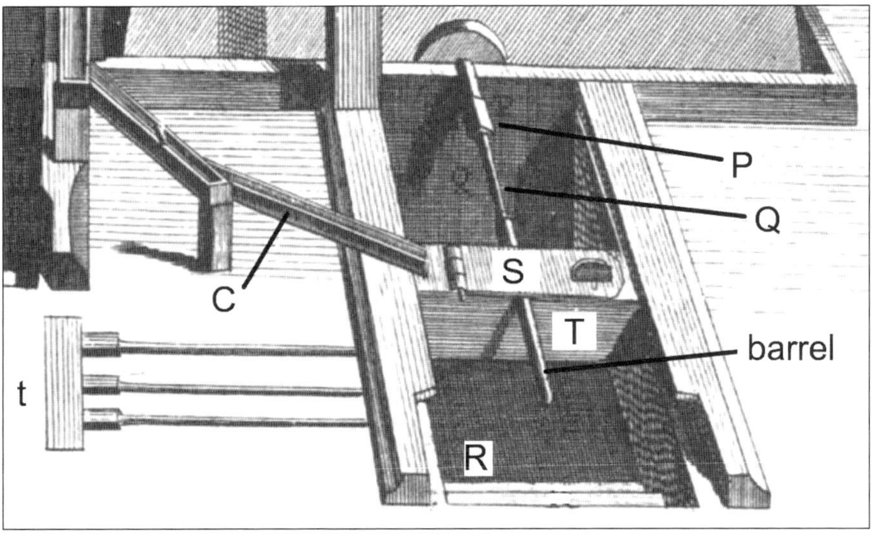

Fig. 3.14. Annotated detail from Diderot (1762) showing a drilling/boring machine.

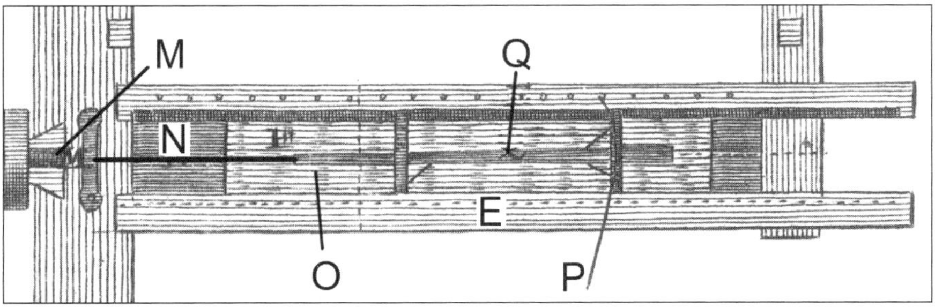

Fig. 3.15. Annotated detail from Plate 2 of Diderot (1779) showing a plan view of the drilling machine.

components forming the bed of the machine, *E*. The drill, *N*, is fitted into a *moussle, M*, [taken to mean a socket on the spindle nose of a powered headstock]. At *P*, a small crowbar was used to force the carriage with the barrel towards or away from the drill.

From the detail provided by Cotty, the drilling operation either embrittled or hardened the iron so that before the next operation it had to be annealed. It also indicates that drilling up to this point used only short drills, requiring the barrel to be reversed so that it could be drilled from both ends: *The iron having been embrittled by the preceding operation, the barrel is annealed in a white wood fire to soften it, and when it has been drawn to the eye,* [straightened by eye?] *the driller uses three or four boring bits . . . to erase the marks of drills, and burnish it internally and give it its right caliber . . . The boring bits differ from the drills only by their length, which is about 40.5 cm while that of the drills is only 21.6 cm. The straightener then straightens the barrel to the cord.*

This straightening of the barrel 'to the cord' is closely similar to that described as being used in Portugal in the early 18th century except that: *The cord is a brass wire stretched by means of a steel bow. The worker passes the wire through the bore and stretches it with the bow to check if it contacts the walls in all its length and in all the directions. By presenting the barrel to the light and turning it over carefully, we see from below if this brass thread applies perfectly to all points of the bore.*[18]

3.4 Finishing the bore in the early 19th century - Tula and the M.1808 musket

In contrast, with regard to many of the machines, the description of operations at Tula by Gamel is full of detail. The barrel drilling machine, Figs 3.16 & 3.17, is reminiscent of those illustrated by Diderot and was considered at Tula to be *of an extremely old-fashioned construction.*[19]

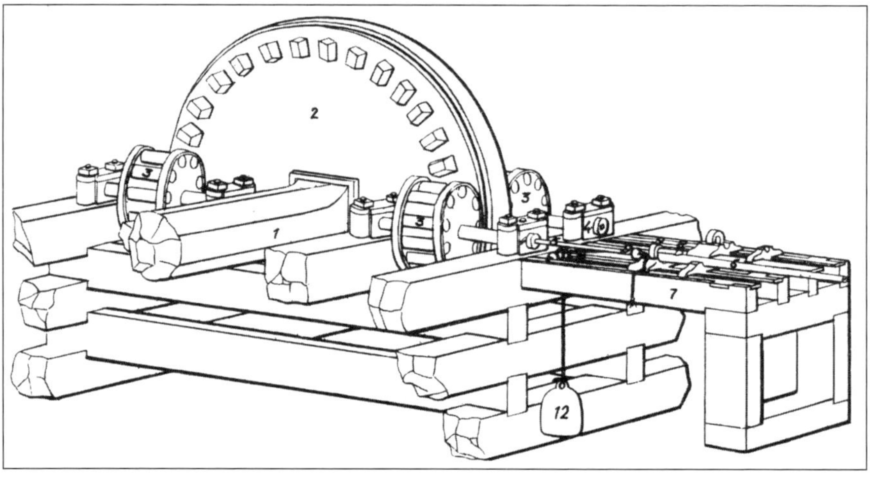

Fig. 3.16 Barrel 'rough drilling' machine in use at Tula in 1826 (Gamel, 1826, Fig V)

The barrel tubes, *6*, were secured to the moving carriage or *chaser*, *9*, by insertion into a hole and fixed by a wooden wedge, *8*, whereas in the machine in *Espingarda* the barrel tube was clamped by a screw. The squared end of the drill, *4*, Fig. 3.18, is fitted into a spindle nose and the barrel tube was fed onto the revolving drill automatically, as in the earlier Portuguese machine, by a weight, *12*, attached to a belt passing over a pulley and attached to the barrel mounted on the cast iron carriage at 10 seen in Fig. 3.17.

Gamel describes the drills used as: *nothing but iron rods whose length is half that of the barrels. The upper part of the drill has a square shape, 4, which fits in the centre of the driving shaft. Its middle part, 14,* [this should read 12 as per Fig. 3.18] *is rounded, whereas the lower part, 15, has a square cross-section on a length of four vershoks* [approximately 7 inches/175 mm]. *The last part is welded with* [spring] *steel, such that the four edges are actually made of steel . . .*[20]

Thus far the drills conform to *drills with square cutting edges made of good steel* described by Cotty. In the same passage Gamel then proceeds to inform us

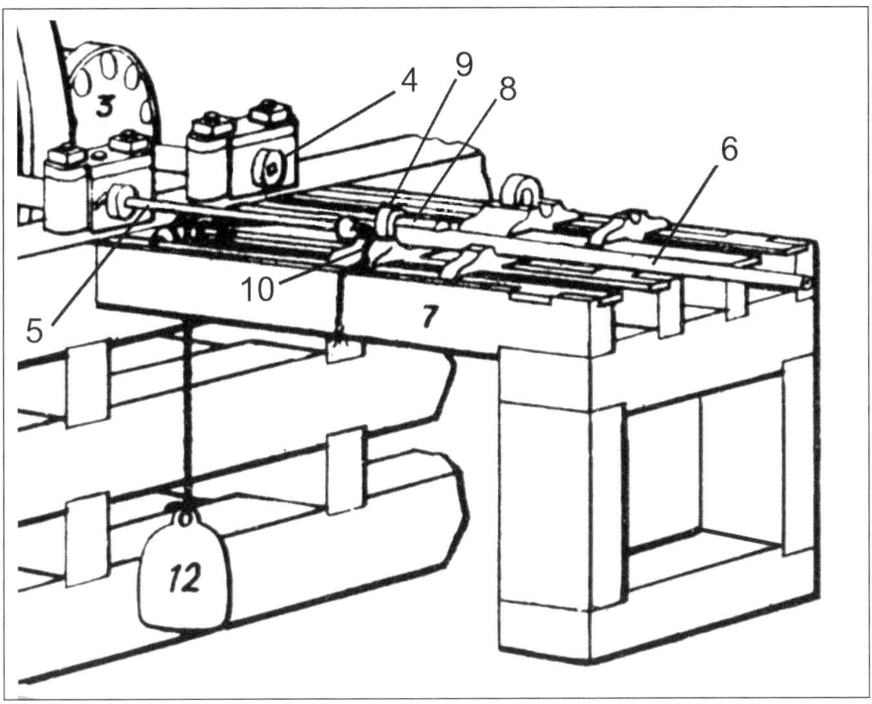

Fig. 3.17. Details of the 1st (rough) drilling machine at Tula (adapted with clearer annotation)

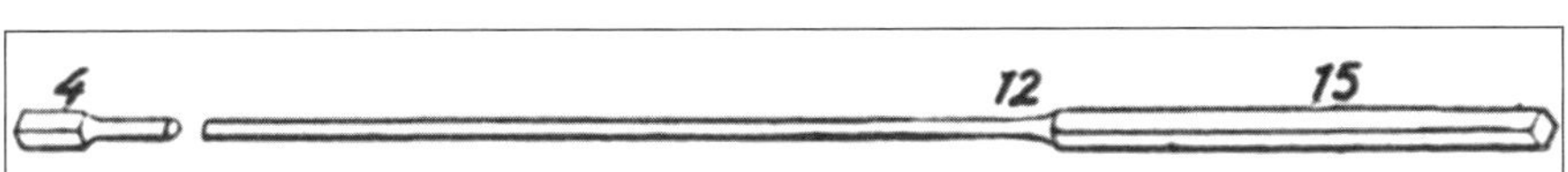

Fig. 3.18. Drill for the first drilling operation at Tula. (from Gamel, 1826, Fig. V)

that: *The drills have twenty-four different sizes which are introduced one after the other, beginning with the smaller size inserted from both ends of the barrel and ending up in its middle part. The drills are lubricated with hempseed oil. The barrel is cooled by means of a wet cloth. A trough with water is put under the barrel for supplying the cloth.*

This again is closely similar to the method used in France. The barrels were drawn off the drill by means of an *extracting rod*. This was like a ramrod; it was inserted into the barrel so that the small button was in contact with the drill, the 'drilling master' then placed the large button, *17* in Fig. 3.19, on his chest and pulled the barrel towards him, forcing the drill out of the opposite end.

The barrels were drilled in this manner until the bore size was six and one-half lines [0.65 inches] and during this process, 1.5 lbs [0.68 kg] of metal were removed. As with the machines illustrated by Diderot, the Tula machines were 'multi-spindle' and each of the three waterwheels used drove twenty-four drills and thus could 'drill' seventy-two barrels at a time. With masters supervising the 'drilling' of two barrels at once and each of the 72 drilling machines capable of producing five rough-drilled barrels per day gave a daily total of 360.

Unfortunately, no figures have been found relating to outputs from the factories in France when Diderot and Cotty were writing but the Tula output takes on the proportions of 'mass-production' compared with gunmaking outlined in *Espingarda Perfeyta* and was undoubtedly the result of military versus private needs.

Even so, Gamel was critical of the technology: *The process of drilling the barrels in the Tulskij factory has not yet reached the fullest possible degree of perfection. According to the present production, the barrels are reamed rather than drilled. Such operations are quite different from each other.*[21] [author's underlining]

Here again we find reference to the process being more akin to reaming than to drilling.

After passing through the rough drilling process the barrel needed to be straightened. According to Gamel, because the drilling process generated heat and the barrel was cooled by water, this caused brittleness in the iron[22] [cf Cotty, 1806, p. 51]. They were annealed and then cleaned with hemp attached to the end of a cleaning rod, followed by straightening using an anvil and a wooden hammer. Then the 'master' viewed the barrel by holding it against the light and striking with an iron hammer on the outside opposite any blisters and cavities that may have existed within. Around three hundred barrels per day were treated in this way by six men.

The second stage of the drilling operation at Tula was different from the first in various ways and used machines installed in 1810.[23] A bank of three double-ended machines, each with two pairs of drilling spindles at each end, were driven by a water wheel through a shaft, *18*. One end of one of these machines with its two pairs of drilling spindles is shown in Fig. 3.20.

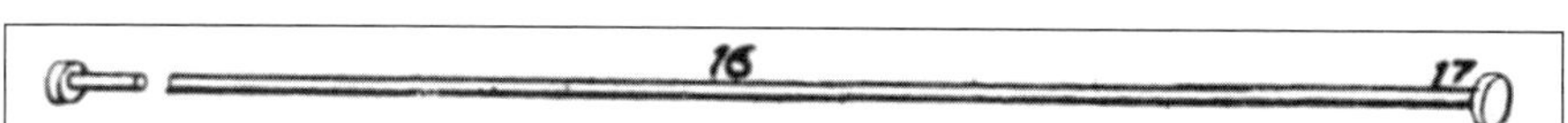

Fig. 3.19. *Extractor Rod* used at Tula. (from Gamel, 1826, Fig. V)

The drive shaft, *18*, was fitted with three double-sided cast-iron 'crown wheels', *19*. Each face of these meshed with a pair of pinions or spur gears in each machine, each of which meshed with another pinion of equal size, *20*. On the outer extremities of the shafts of each pinion was a socket, *21*, into which a drill could be fitted. The report states that: *One master operates on each pair of machines; he is helped by twelve drillers. Twenty-five barrels are drilled per day on each pair of machines.*[24]

The daily total for the bank of three double-ended machines would therefore be seventy-five.

Examining the machine in greater detail, in Fig. 3.21 the drills, *22*, are driven by means of sockets at the ends of the spur gear shafts, *21*. The barrels were fixed by screw clamps on the 'chasers' [carriages], *26*, which slid between iron guides, *28*, fitted to the wooden bed members, *27*.

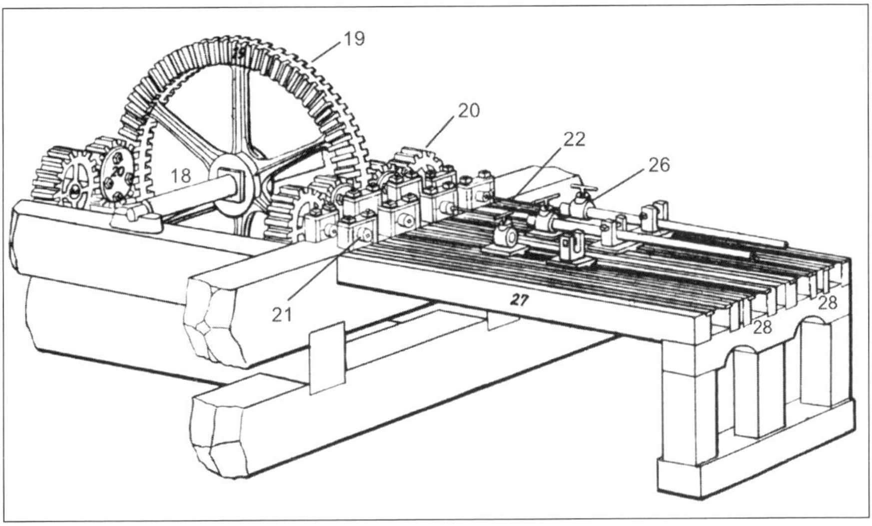

Fig. 3.20. Part of the 'second drilling' machine at Tula with added annotations. (from Gamel, 1826, Fig. V)

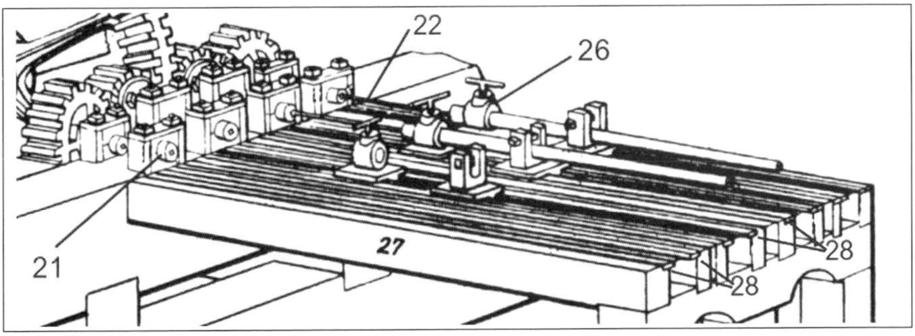

Fig. 3.21. Detail of the drilling machine with added annotations. (from Gamel, 1826, Fig. V)

Unlike the drills for rough drilling shown in Fig. 3.18, this second drilling operation used drills that had a radially projecting peg, *23*, close to the end of the shaft, Fig. 3.22.

The round end of the drill shaft fitted into a central hole on the drive shaft and the peg fitted into a slot in the 'T' shaped projection, *24*, Fig. 3.23, in a similar manner to the 'drive-dog' and 'catch-plate' on a conventional lathe.

Because each pair of spur gears rotated in opposite directions, the slot for the peg on the drill was double-sided, the slot being engaged depending upon the direction of rotation. This was part of a very ingenious system in which the drills themselves were self-feeding. What is not apparent from the illustrations in the Tula report is that not only were the square 'working' ends of the drills 'steel plated on both sides', but they were also *serrated along their corners.* In a footnote the comment is made: *It seems possible to drill the barrels by sharp tools, setting them in such a way that the cutting tool is gradually screwed into the barrel from the surface* [muzzle?] *to the bottom* [breech?]. *Moreover, water should flow continuously in the barrel to cool the cutting tool and wash away the cut chips.*[25]

It is presumed that these serrations must have been arranged in a regular helical pattern, as in a screw, similar to that shown in Fig. 3.24, because when the barrel, clamped in the carriage, was pushed so that the rotating drill, entered the breech end, it: *behaves like a screw and pulls the barrel by the serrations in its direction. When the drill goes right through, the Master removes the barrel without drawing the drill.*

The drill, with the barrel still on it, was then inserted, *into the centre of the shaft of the gear which is rotating in the opposite direction. Therefore, the drill is self-drawn from the barrel.*[26]

This seems something of a conundrum, but without knowing the exact geometry of the drills, or more correctly, their serrations, it is difficult to be certain. The drills

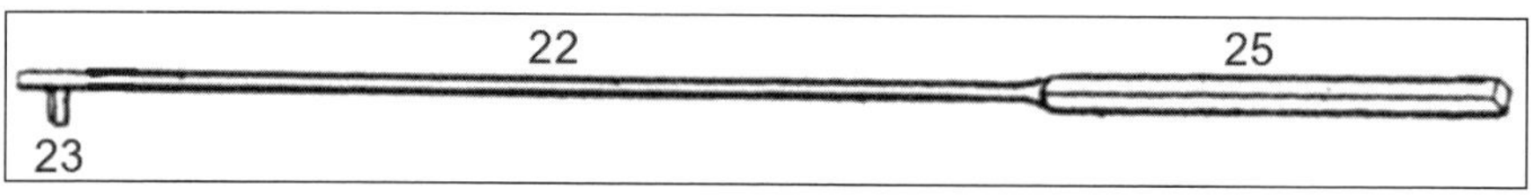

Fig. 3.22. The drill for the second drilling operation. (from Gamel, 1826, Fig. V)

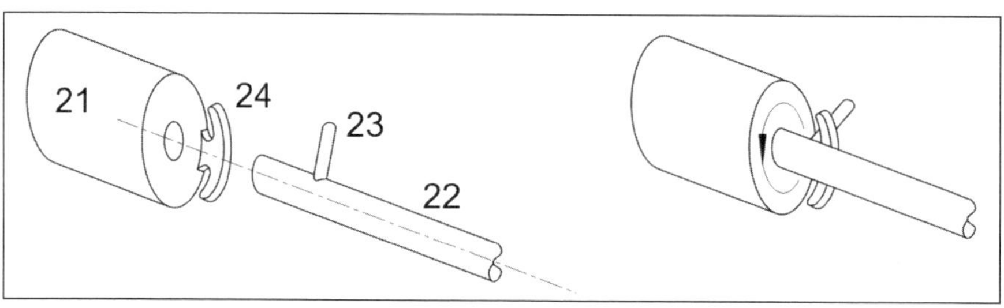

Fig. 3.23. Sketch showing how the drill in Fig. 22 engages with the drive-spindle nose for the second drilling operation. The peg, 23, on the drill shaft engages with one or other of the forks on the spindle nose depending upon direction of rotation. (© P. Smithurst)

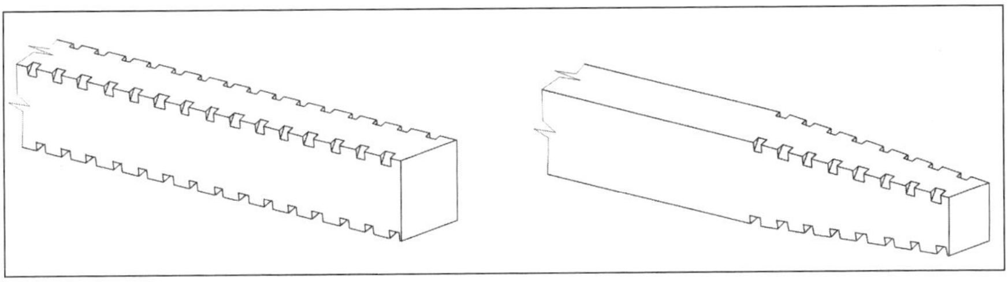

Fig. 3.24. Sketches of how the *serrations* on the end of the 2nd drilling bit may have appeared, arranged in a helical pattern so that the bit acted as a square screw. Although not stated, the drill may have been slightly tapered at the nose as shown on the right (exaggerated) to enable it to 'start' the cutting operation more easily, the remainder being plain. (© P. Smithurst)

were there to remove metal, not cut threads - Gamel tells us, for instance, that in the second drilling operation, half a pound (0.18 kg) of chips was produced. But, with an ordinary threading tap, metal is only removed to match the geometry of the teeth, creating a helical groove. If it then removed the metal between the grooves, it would be a pointless exercise! But that is exactly what is needed in drilling the barrel. If the serrations on the drill only occupied the leading portion of the drill, then one could envisage a situation where they acted to pull the drill into and through the barrel, while the un-serrated faces which followed removed – 'reamed out' – the threads thus formed. It could be argued that in such a case, the drill would not need to be 'screwed out'. However, it may indeed have cut a thread all the way through because after stating that the drill is *self-drawn from the barrel*, the passage continues: *A wooden strip is fitted to one of the four sides of the drill so that it goes through the barrel closely and straightforwardly. Hempseed oil is also used here for lubrication.*

This conundrum may, of course, be simply due to mistranslation but if so, careful editing should have picked it up.

On these machines the barrels were drilled to a calibre of six and three quarter lines [0.675 inches]. Thus, the amount of metal being removed corresponded to an increase of 0.025 inches on diameter, or 0.0125 increase in radius, which in turn meant the depth of the serrations could only be very small, in which case it seems surprising that the threads were not 'stripped' in the process. The barrel was straightened once again, each of the eight workers straightening 50 barrels per day.

The third and final drilling operation *for bringing it to the final seven-line calibre* [0.7 inches, an almost standard size for muskets], in which a further increase of bore diameter of 0.025 inches resulted, was carried out on yet another machine, Fig. 3.25, developed in 1825.

This new machine differed in having all its principal components being made of cast iron instead of wood, resembling 'modern' machines in that respect. It was therefore more rigid and capable of greater accuracy.[27] The machine was also technically more advanced in having a reversing mechanism utilising a simple sliding 'dog-clutch' keyed to the main shaft and engaging with 'dogs' on the two wheels *36* and *33* and fitted with a forked clutch lever, *35*. For transition from

Fig. 3.25. Machine for the final drilling operation at Tula, this time built of cast iron. (Gamel, 1826, Fig. V)

'forward' to 'reverse' it would need also to ideally have had a 'neutral' position to change the drive direction, and this would have simplified fitting and removal of drills and barrels.

Some further confusion arises in reading the report at this point. So far, no mention has been encountered regarding any external machining of the barrel. However, in a footnote to the section dealing with the final drilling operation it is stated: *In the previous machining operations of the barrel, which is mentioned above, when the barrel was drilled for the last time after turning its external surface*, [author's underlining][28] which indicates some turning of the outer surface took place before the final drilling operation.

However, since such an action is not included in the main body of the book, it is difficult to offer any worthwhile comment. The drill used in this final drilling operation is not illustrated but seems to be of similar construction in most respects to that used in the second drilling operation, Figs. 3.20, 3.22. It was fitted into the nose of the driving shaft, Fig. 3.26, and is driven by peg engaging with a fork on the shaft nose.

The comment, *a semi-circular wooden strip is put on one side of the 'blade' of the drill and is lubricated with hempseed oil* suggests that the working end of the drill itself was 'flat', rather than of square section.

Nevertheless, the drill was serrated in the same fashion: *The drill, inserted in the barrel and rotated by the wheel (33)* (see Fig. 3.25), *screws the barrel on itself by means of the small teeth cut on it.*

The 'working end' of this drill might therefore be something like that in Fig. 3.27 and approximates in some respects to later 'gun drills' which were all-steel, semi-

Right: Fig. 3.26. The spindle nose of the third drilling machine. (from Gamel, 1826, Fig. V)

Below: Fig. 3.27. Sketch showing how the working end of the third drill may have looked to allow the fitting of a not-quite semi-circular wooden strip. (© P. Smithurst)

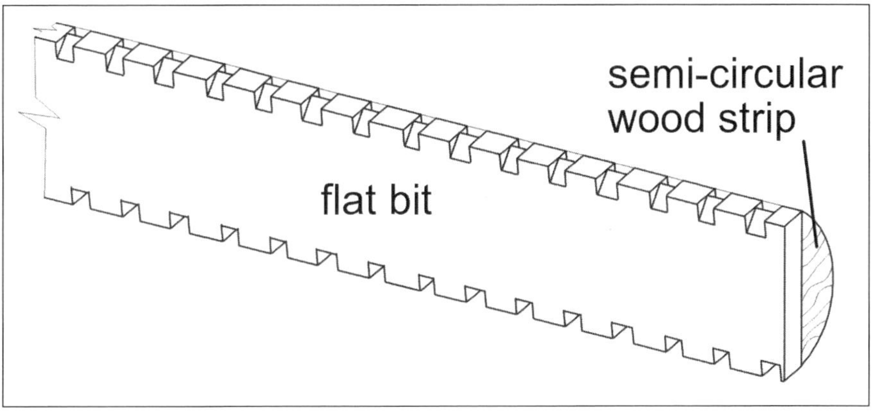

circular, but without serrations, and used for deep hole drilling. By reversing the machine, the drill was withdrawn from the barrel and then re-inserted but with a thin strip of 'bast'[29] placed under the wooden strip already in place. This caused slightly more metal to be removed.

At each stage of drilling the barrel was checked for any hollows in the bore and, if any were found, it was sent to the straighteners to have them 'filled with iron'. Exactly what this means, or how it was accomplished, is not stated. However, in a later article describing operations at the Springfield Armory in Massachusetts, in dealing with what are referred to as 'cinder holes', created by cutting through small bubbles which had formed within wrought iron bars at the time of their manufacture, or pulling out small cinders, the following comment is made: *Sometimes the iron may be forced in, by a blow upon the outside, sufficiently to enable the workman to bore the cinder hole out entirely. This course is always adopted where the thickness of the iron will allow it. . .* [author's underlining][30]

The small size of a cinder hole is different from what might be termed a 'hollow', but the term 'filled with iron' seems likely to mean simply hammering the outside of the barrel so that it bulges sufficiently inwards at the right place so that when re-drilled, the fault was eliminated. The drilling operation was repeated about twenty times until: *the cylindrical gauge (having a precise diameter of seven lines), inserted into the vertically set barrel, passes freely along the whole length of the barrel and entirely fitting to it everywhere* – a rare instance of the mention of a gauge.[31] This is virtually identical to the system described as used in Portugal in the early 18th century and is another instance of the early adoption of gauging.

The final operation in forming the bore at Tula was 'honing' in which any marks left by the drilling processes were removed. The honing machine, Fig. 3.28, was devised and installed at Tula by a Mr Zahava in 1810 and replaced hand-tool methods, utilising a complex system to provide a reciprocating motion to the 'hones'.[32]

The machine was driven by a water wheel via the connecting rod, *1*, which was coupled to a rocking beam, *2*, at the opposite end of which was a sector arm, *3*. Two chains, attached to each end of the sector arm, were tightly wrapped in opposite directions around a drum. Thus, as the beam rocked, these two chains caused the drum to oscillate about its axis, undergoing a partial rotation in each direction.

On the same shaft as this drum was a second drum mounted centrally beneath the machine. This was fitted with two ropes, *7*, in the same manner as the chains on the first drum, and the other end of each rope was connected through tensioning bolts to opposite ends of a bed, *8*, which slid between two guides. Through this arrangement, then, movement of the beam resulted in a reciprocating motion of the bed, *8*, between its guides. Sliding between the same guides were two small subsidiary beds or 'chasers', *10*, Fig. 3.30.

Each of the chasers had a simple spring-loaded hook arrangement, *9*, to link them to opposite ends of the main bed, *8*. Through the centre of the chaser was a longitudinal hole into which was fitted a pin, *1*. Secured to this pin was a ratchet wheel, *22*, engaging with a pawl with a long arm, *24*, and held in position by two spring-loaded detents, *23*. The forward end of the bolt, *1*, was formed into a forked knuckle into which was fitted the eye, *16*, at the end of the hone rod, *12*, and was secured by a bolt, *20*.

Unlike our present concept of honing using fine grindstones, at Tula a pair of small segmental files were used and it would seem appropriate for removing the, albeit fine, spiral grooves cut in the bore by the drilling operations. These were fitted into a dovetail slot in each of two arms, *14*, formed by slitting the end of the long honing bar, *12*. The separation of the arms carrying the file inserts was regulated by the screw, *15*, so that the surfaces of the files could be adjusted to correspond with the bore size, Fig. 3.31.

As files they could only cut in one direction and they did not encompass all of the bore so, after passing down the bore to perform their cutting stroke, they were withdrawn and automatically rotated slightly before passing down the bore again, thus treating the whole of the walls of the bore.

The reciprocating motion of the bed, *8*, has already been explained. This motion of the bed was transferred to the *chasers* through their linkage to it, and through them, to the hones themselves. The necessary indexing, or stepwise rotation of the hone, was accomplished by the mechanism fitted to the chaser and to which the hone was connected. Towards the end of the rearward stroke of the chasers, the extended arm of the pawl, *24*, encountered an inclined plane, *25*, causing the ratchet wheel to rotate the hone into position for its next stroke.

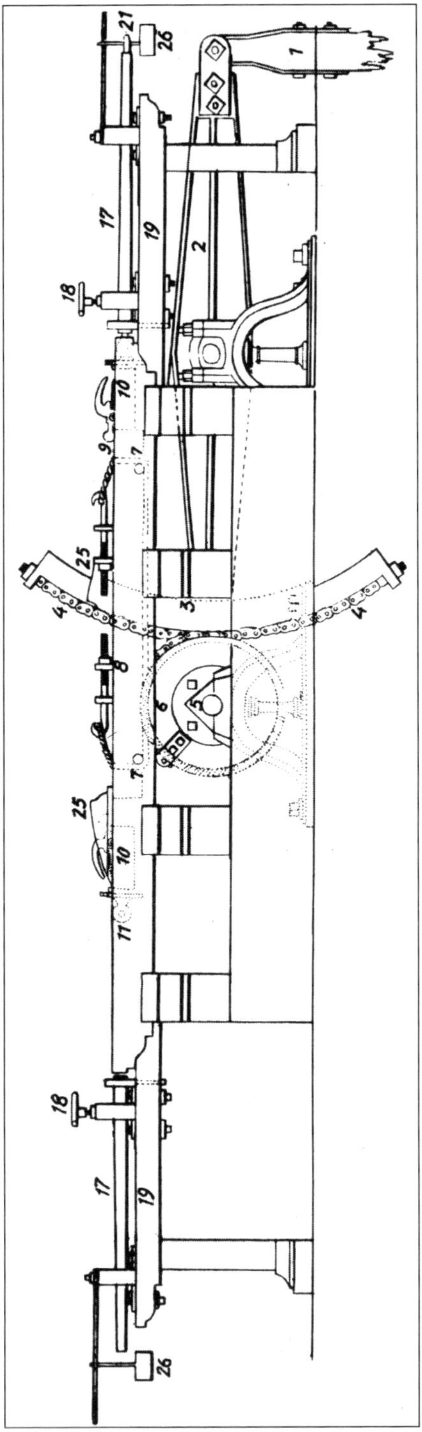

Fig. 3.28. The 'honing' machine used at Tula. Bearing in mind a musket barrel was just over 1 metre long, this machine had an overall length approaching 6 metres. (Gamel, 1826, Fig. VI)

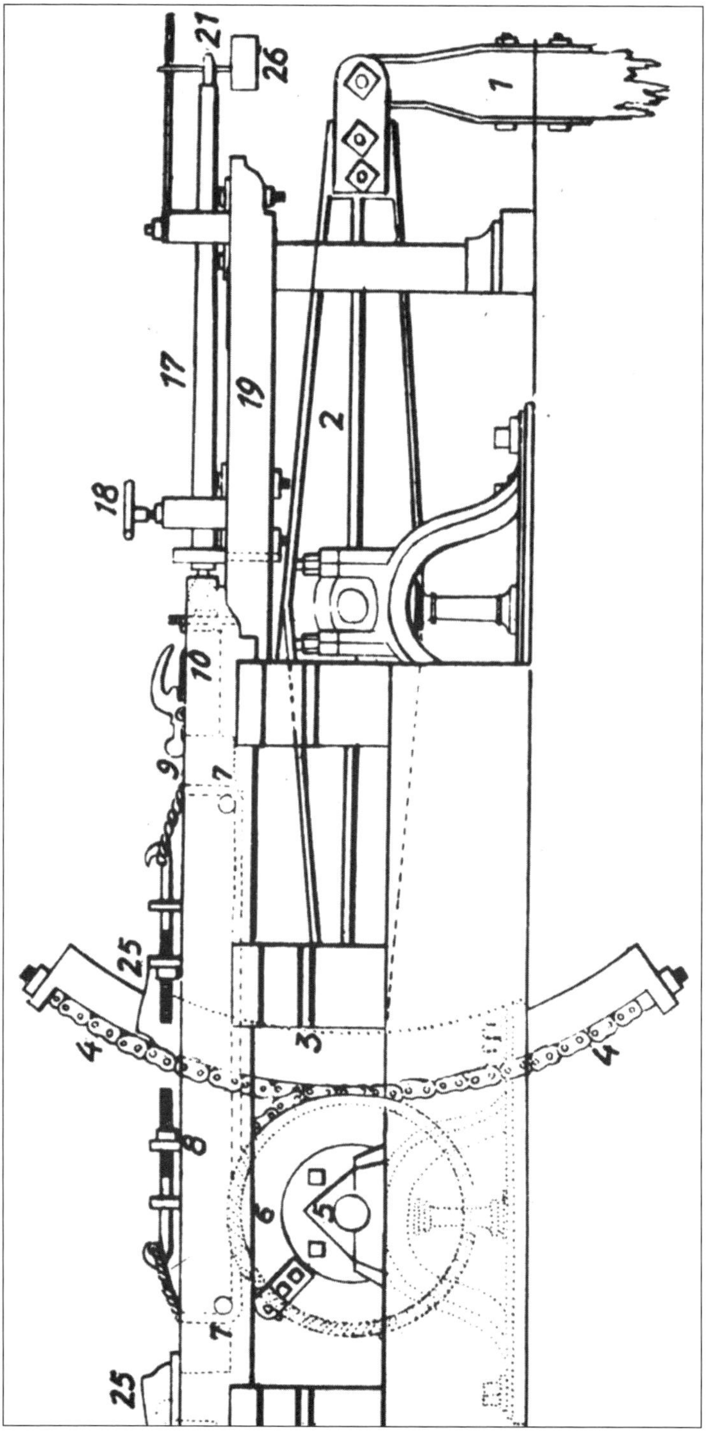

Fig. 3.29. The main elements of the 'honing' machine in more detail.

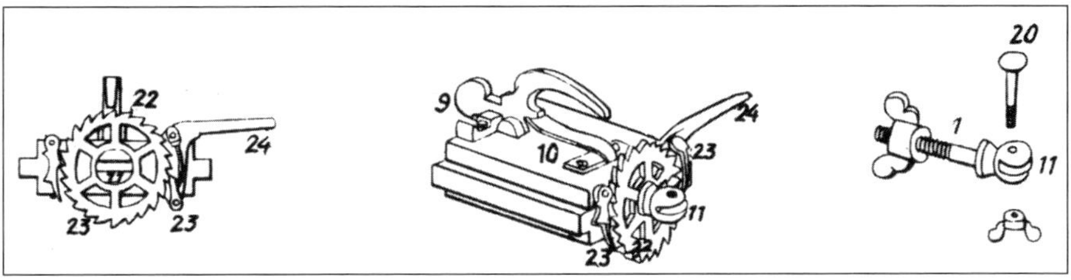

Fig. 3.30. Details of the 'chaser', showing the ratchet indexing mechanism and the means of connecting it with the 'hone'. (from Gamel, 1826, Fig. VI)

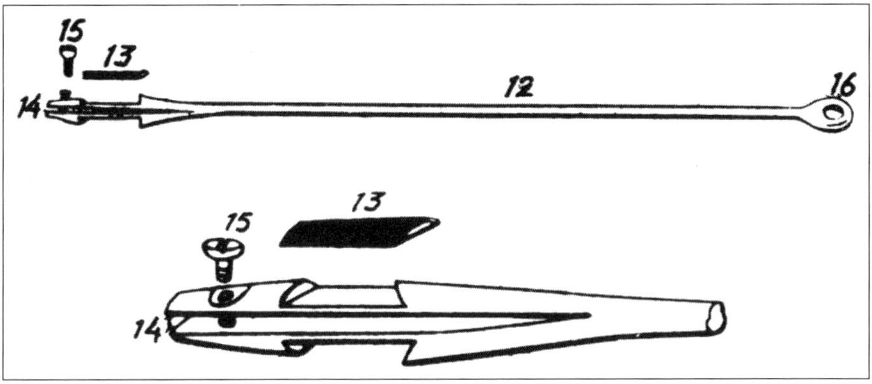

Fig. 3.31. Details of the 'hone', showing the eye, 16, by which it is connected to the 'chaser', and the forked ends of the rod, 14, carrying the file segments, 13, and whose separation is adjusted by the screw, 15. (from Gamel, 1826, Fig. VI)

Fig. 3.32. The indexing mechanism was actuated by the arms of the pawl, 24, riding up the incline, 25.

This honing machine represents a completely different technology for finishing the bore to that being used in France only a few years previously and represents a high degree of innovation in the Russian approach to gun manufacture at that time. At this point, the work on the bore of the barrel was temporarily halted, pending work on the exterior.

3.5 Finishing the bore in early 19th century America

Although America adopted at least variants, if not the entirety, of the 1777 musket, there seems to be little recorded detail of the processes employed. What does exist provides an interesting comparison with the technology employed at Tula in this period. In 1814 an important patent was granted to Pettibone in the United States. It shows a spiral-fluted drill or boring bit, Fig. 3.33.

However, detailed study and comparison is not possible due to the loss of the original Pettibone patent documentation in the Patent Office fire of 1836, but from a copy of the illustration which survives it suggests it is probably the first example of a spiral fluted drill of any kind although it could only work on a pre-existing hole, unlike the later Jennings and Morse drills for wood and metal, and therefore acted more as a reamer.

Fitch[33] suggests that an improved ‘nut-boring’ (rough boring) machine using an ‘auger’ was introduced at Harper’s Ferry National Armory by James Green in 1817. He probably meant James Greer who, according to Smith,[34] apart from being the first machinist at Harper’s Ferry, claimed to have invented the screw auger for barrel drilling in 1797 and first used it at Robert McCormick’s Globe Mills Armory a year later. Smith also suggests it may have been used in England even earlier although no evidence has been found for this, and states that others, such as William Holmes of Springfield Armory and Lemuel Pomeroy of Pittsfield, Massachusetts, also laid claim to it. But despite various patent litigations and the possibility of a form of parallel evolution, Pettibone’s is the only one we have real evidence and details of.

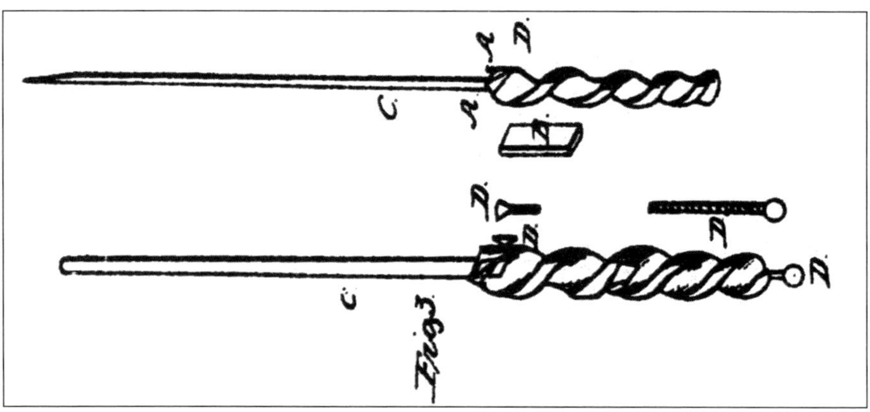

Fig. 3.33. Details of spiral fluted drills from Pettibone’s patent.

When it comes to early barrel drilling/boring machines, we have less information. Pettibone's patent also includes a drawing of a machine, Fig. 3.34, in which the central drive for the drill is flanked by two gears driving shafts fitted with worms that engage with teeth on a sliding carriage and feed the barrel onto the drill. No doubt two worms were used to engage with each side of the carriage to prevent 'crabbing' but without the accompanying text further attempts at analysis would be fruitless.

Nevertheless, by 1821, Smith[35] also notes that Springfield Armory had taken the innovatory step and adopted what is essentially the boring machine shown in Pettibone's patent, though how far they may have been copying Harper's Ferry is uncertain. However, this technology would seem to lag behind the mechanical sophistication being displayed at Tula.

As in previous cases, rough-boring was followed by smooth-boring and involved the passage of a succession of what by this time were traditional *rectangularly shaped cutting bits*[36] through the barrel. This was then followed by fitting the breech plug. The chamber portion was reamed with a two-inch long reamer and then set vertically in a vice and the chamber tapped. Two 'conically shaped' taps of increasing diameter were screwed into the chamber by hand, followed by a cylindrical tap with the same diameter as the breech plug. This process of using three taps – a taper tap, 2nd tap and 'plug' tap - is still used today. It was followed by removing the 'burrs' raised by tapping on the mouth of the breech and the drilling of the vent. Here again, Tula seems more advanced in its approach to this and here, as in Portugal and France, 'breeching' was performed after finishing the outside of the barrel.

Despite Pettibone [or Greer?] having developed new barrel drilling tools at Harper's Ferry Armory, most barrel drilling tended to use the traditional tools. At the Springfield Armory in Massachusetts, according to one article, boring was still carried out: *By means of certain tools which are called augers, though they bear very little resemblance to the carpenter's instrument so named. These augers are short square bars of steel.* A similar phrase is used to describe barrel boring at the Norwich Armouries in 1864, creating ambiguity in the use of the term 'auger'. This article also offers additional support for the contention that at Springfield, and no doubt elsewhere, internal and external work is often 'intermingled' but, frustratingly, fails to elaborate: *In fact the borings to which the barrel are subject, alternate in more*

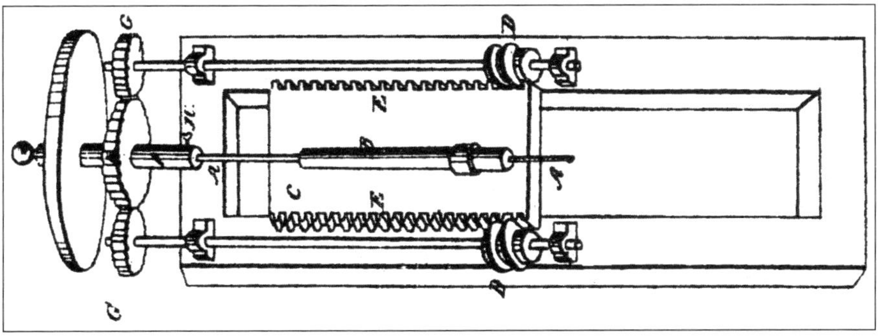

Fig. 3.34. Pettibone's barrel drilling machine.

instances than one with other operations, the whole forming a system far too nice and complicated to be described fully within the limits to which we are necessarily confined in such an article as this. It is a general principle however that the inside work is kept always in advance of the outside, as it is the custom with all machinists and turners to adopt the rule that is so indispensable and excellent in morals, namely, to make all right first within, and then to attend to the exterior.[37]

One innovation in American practice which Fitch reports,[38] though does not suggest when it was first used and for which no evidence of its use anywhere else has been found, is that the 'bits' were placed under tension during use. This is deduced from his comment that the drill shaft was first passed through the barrel, followed by connecting it to the driving shaft and the barrel fed onto the drill in the opposite direction. This was to help keep them, and therefore the bore they produced, straight. A barrel drilling machine being used at Springfield in 1878, illustrated by Benton[39] and again by Fitch, Fig. 3.35, incorporated this technique, but unfortunately these details are not apparent. However, it still relied on being hand-fed by means of a handwheel and rack and pinion gearing, which seems

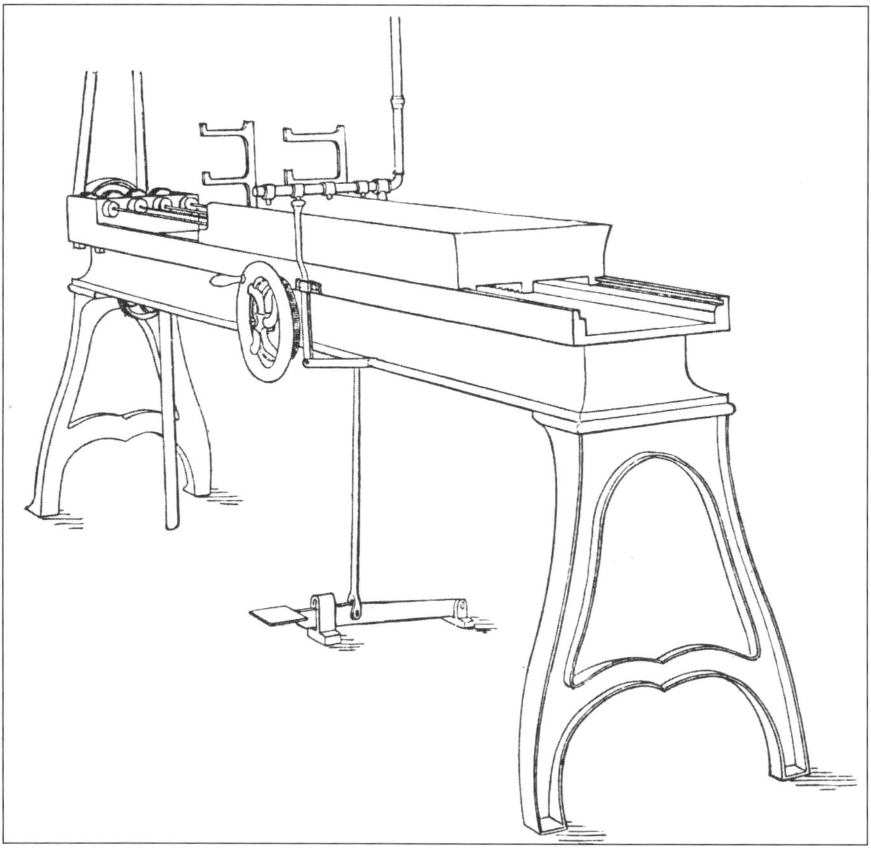

Fig. 3.35. Machine at Springfield using drills placed under tension. (Fitch, 1882, Fig. 7)

something of an anachronism in the light of his statement that *automatic feed barrel boring machines were in use at Whitneyville prior to 1840,* but he offers no indication as to how these functioned.[40]

3.6 Finishing the bore in early 19th century Britain

In Britain, barrel boring largely continued on traditional lines using square 'bits' within both the 'sporting gun' trade and in military barrel manufacture in the private trades and at Enfield. In terms of sporting guns, this method continued well into the late 19th century and was also being practiced in the early years of the 20th century. Thus, even such a prolific, and predominantly sporting gunmaker as W. W. Greener stuck to the traditional methods of boring barrels, which, from what he describes[41], were almost identical to those used in Portugal in the early 18th century and seem antiquated when compared with the machines used at Tula in 1826. The machine illustrated by Greener, Fig. 3.36, was actually used for fine boring but is similar to that used for the initial process of rough boring except in some of the details, and its similarity to its forbears is immediately apparent. The barrels were held in a carriage which can traverse the whole length of the bed of the machine.

The drill in this process, as in others, was *a straight square rod of steel, slightly tapered at the point, and about five feet long* and revolved at about 300 rpm. It might be inferred from this that the drill is square throughout its length but, from an illustration of the drill used for 'fine boring', it is presumed that this one, like others encountered, does have a reduced-size round shank, as in Fig. 3.37, but the 'tail-end' of the rough-boring drill was flattened and had a hole drilled through

Fig. 3.36. Barrel (fine) boring machine. (Greener, 1884, p. 272)

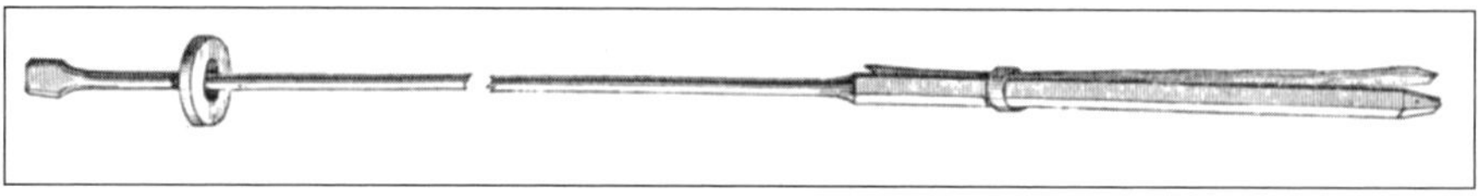

Fig. 3.37. A fine-boring bit with a spill and any paper packing strips held in place by a ferrule. (Greener, 1884, Fig. 191, p. 273)

into which a pin was fitted to secure the bit to the machine spindle. An iron rack was fitted to the bed and a small crowbar engaged in the rack to force the barrel on its carriage onto a drilling bit of suitable size so that the bit passed through the barrel. The barrel was taken off the bit, the next size of bit selected, and the process repeated until all the 'fire scale' [flakes of iron oxide produced during heating and welding] had been removed and the barrel was of the correct bore size.

If all the fire scale had not been removed, due to some small 'hollow' in the barrel, it was returned to the welder who, as noted elsewhere, heated it and carefully hammered the 'hollow' portion on the outside of the barrel which was then re-bored. During rough boring a stream of cold water was directed onto the barrel [again like methods used in France and at Tula]. The process did not apparently require skilled labour, unlike the 'fine boring' stage.[42]

The fine boring machine, Fig. 3.36, was equipped with a weight and chain used to feed the barrel onto the boring bit, reminiscent of the technique used in Portugal (Figs. 3.4, 3.5) and in the rough boring stage at Tula, Fig. 3.16. A similar square drilling bit, revolving at only half the speed or less of the bit in the rough boring operation, was used but it had significant differences. It had only one sharpened cutting edge or corner, the adjacent one being rounded to act as a burnisher whilst the remaining two were kept out of contact with the bore by its being packed to fit the bore by a sliver of wood and strips of paper called *liners*. This sliver of wood, commonly referred to as a *spill* – equivalent to the *reed* encountered earlier in Portuguese practice – gave rise to this fine boring process becoming known as 'spill boring'.[43] A fine-boring bit illustrated by Greener is shown in Fig. 3.37. A 'spill' and any 'liners' are held on the square head of the bit by a ferrule which was pushed off as the bit entered the bore.

The nature of the bit and its action in conjunction with the spill may be made clearer in the schematic cross-section of a barrel being bored, Fig. 3.38.

It will be noticed that with the spill in place the centre of the bit, having had a corner rounded, is offset horizontally and vertically from the centre of the bore due to alteration of the geometry to bring the cutting edge and the rounded burnishing edge into contact so that they can perform their respective functions. Theoretically, this would suggest that unless the drill can 'float' to a small degree in its socket in the driving spindle, the square of the bit would be canted so that it would be at a small angle to the bore axis. Nevertheless, this method was able to achieve near perfect results, otherwise it would not have persisted for so long.

Greener also notes that barrels had their bores finished by 'lapping'. In many ways it is more in keeping with our current concept of honing than that practised at Tula. It was really a polishing process aimed at producing a bore with a mirror-

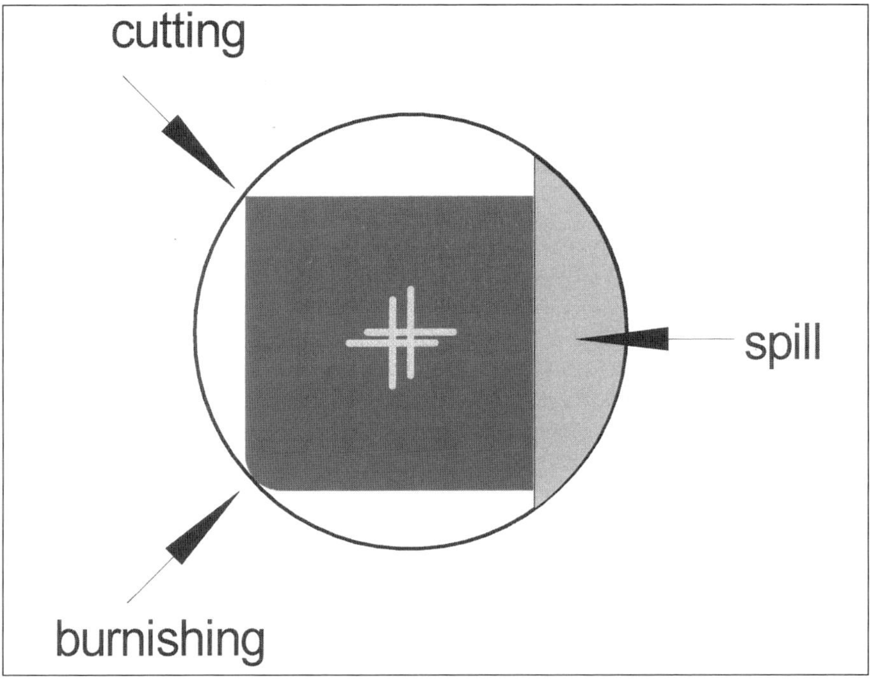

Fig. 3.38. Exaggerated cross-section through a barrel with the bit and spill in place. (© P. Smithurst)

finish, rather than adjusting the bore size within fine limits, though that would, of course, be another outcome. The lap detailed by Greener[44] consisted of an iron rod upon which was cast a lead *case* of the same diameter as the bore to be treated. This seems to imply the whole length of the rod had this lead covering, rather than just the 'head' portion. The machine used was similar to a lathe and the lap was fixed into the headstock spindle which revolved at 650 revolutions per minute; during use the lap was constantly supplied with a mixture of emery and oil. The barrel was fixed on a carriage and having inserted the lap and set it in motion, the barrel on its carriage was moved backwards and forwards on the lap, removing any slight irregularities left from the boring processes and polishing it as fine as possible *which is necessary if first class regular shooting is to be obtained.* It also rendered the barrel easier to clean and less likely to 'lead' or 'foul', i.e., suffer from a build-up of minute deposits of lead and combustion products from the powder.

The 'private' gunmaking trades probably retained 'traditional' methods partly due to the fine specialisations, and where self-employed 'barrel-borers' constituted a distinct and separate trade within the gun trade as a whole, and where to adopt new technology meant considerable capital outlay on new machines. Add to this lost time, and quite likely a few spoiled barrels while becoming familiar with the new machines, before production could begin again, was a financial burden independent craftsmen could ill-afford.

Chapter 4

Finishing the Barrel Externally

4.1 Introduction

The external finishing of the barrel encompassed a variety of tasks. The making and fitting of the breech plugs usually occurred at this point; any sights to be used were fitted. For sporting guns, though the barrels may have been finished all over, the polishing and etching to bring out the patterns of *Damascus* barrels was carried out. Since military flintlock barrels were a more or less plain tube, they could be fully turned on a lathe, or simply ground. Also, most military muskets had bright barrels, so they needed polishing to remove grinding or turning marks.

For many of those engaged in the production of military guns, hand methods were combined with simple mechanical methods as they were developed. For military barrels there was also a gradual transition to increased mechanisation as a means of speeding up the process and reducing costs. Machines also played an important part throughout in the quest for closer uniformity and interchangeability.

4.2 External finishing in early 18th century Portugal

At the time *Espingarda Perfeyta* appeared, some thought had already been given to easing the burden of hand-filing. Thus, in Portugal the barrels were finished in two basic ways, either by the use of files alone or the use of files *and* a lathe, and these processes were applied to part-octagonal and part-round, and to plain round barrels. However, we are once more confronted by ambiguity, possibly arising from errors in translation of unfamiliar archaic language and terminology. *Espingarda Perfeyta* bases most of its descriptions on sporting gun barrels with octagonal breeches but again, many of the same principles apply to plain round barrels.

It would appear that throughout the preparation of the outside of the barrel, a 'compass', a form of gauge, was used. It had one leg which, aided by springs, was a close fit in the bore and the second leg was fitted with screws which could be set against an existing barrel deemed perfect enough to indicate the correct 'diameters' of the octagon faces or the round portion. Such a 'compass' is shown being used in Fig. 4.1.

Fig. 4.1. Left: use of the 'compass'; Right: detail - being used to determine and check outside diameters at various points of the barrel. (Fiosconi and Guserio, 1718, p. 94)

In preparation for it being turned on the lathe: *there shall be placed outside it a wooden reel, on which there is to go a cord, this reel just in the middle of the octagonal part, and well and straight.*[1]

This wooden reel with its cord in place as a means of rotating the barrel can be clearly seen in the illustration in Fig. 4.2. The muzzle and breech ends of the bore of the barrel were fitted on conical 'centres', mounted in what we would refer to as the headstock and tailstock of the lathe, liberally oiled, and driven by means of a hand-turned flywheel cum pulley.

In treating plain round barrels, we are told: *there are three methods of filing round barrels, for some are all smooth, having only two octagon faces a little more than four fingers' length, one on the side of the touch-hole and one on the side opposite. These octagon faces are made in all-round barrels, and without them they are imperfect.*[2]

These are comparable with musket barrels, the 'octagon faces' creating the flats on each side of the barrel at the breech end.

At this point, the authors of *Espingarda* appear to abandon logic by instructing the barrel maker:

> *it will be seen which part has to be filed, very straight octagon faces then being made, and governed by the rule. And after being made truly octagonal, it shall be given sixteen faces and then round . . . After working it lengthwise with the cepilhos,*[3] *it shall be taken to the*

Fig. 4.2. A barrel turning lathe and tooling (Fiosconi and Guserio, 1718, p. 80)

> *turning lathe, there to be worked with the gouging chisels in such a way that these . . . leave it truly round.*

It starts as a round barrel and ends as a round barrel with two flats on opposite sides, which seems a very laborious way of effecting the transformation exhibited in Fig. 4.3.

It would appear to have been easier to go directly from the forged and bored barrel, *1*, to the barrel with flats, *4*, by filing the flats and then turning in the lathe.

The barrel was finished to a gauge, Fig. 4.4. While this was for use with barrels having an octagonal breech, one *vara* (43 inches/1100 mm) in length with a bore of 14 *adarmes* (0.786 - 0.804 inches/19.9 - 21.3 mm) and weighing three and one half *arreteis* (3.5 lbs/1.6 kg), with deeper openings, these could equally have served

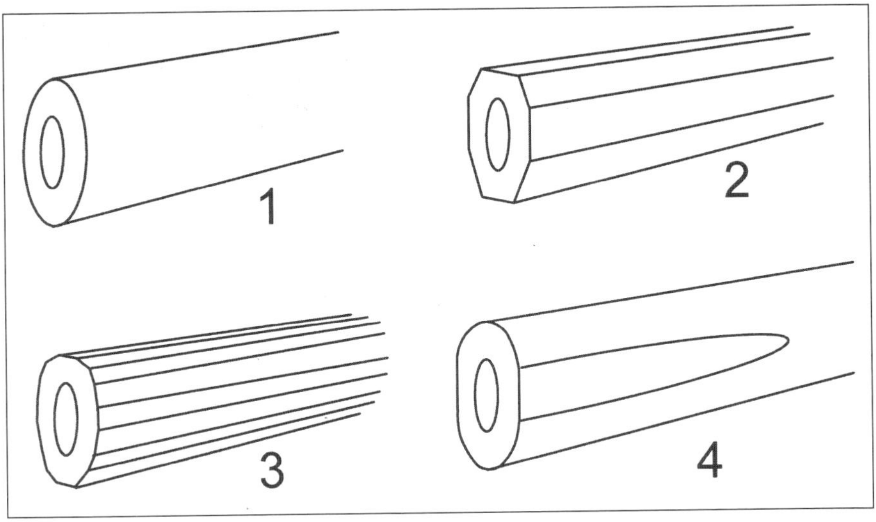

Fig. 4.3. The transformations implied in *Espengarda Perfeyta* (© P. Smithurst)

for round barrels. The openings in the gauges reflect what were considered the critical sizes to achieve those objectives and despite being used for an octagonally breeched gun, it is nevertheless instructive to consider its application.

The openings on the gauge are designated and subdivided into units of *grains*, the definition of which has thus far proved elusive. It cannot be taken to be the same as the English *grain* used to define the inch in the reign of Edward II as being *three grains of barley, dry and sound, placed end to end lengthwise*[4] since that would create a ludicrously large diameter for such a barrel. However, the fifth opening is designated for the smallest diameter which is applied *one couto* (4 inches/100 mm) *behind the muzzle, since this is the most slender part existing in it*[5], (i.e., a 'swamped' barrel typical of many sporting guns of the period) and is set at 8 *grains*. Assuming the bore size as approximating to 0.8 inches, we might take the outer diameter at this point to be at least 0.9 inches, thus making the *grain* approximately 0.11 inches.

The first opening, *1*, is the thickness of the barrel at the extremity of the breech where the breech plug was fitted; if we take this to be 1.6 inches, with a bore size of approximately 0.8 inches, this suggests a wall thickness of 0.4 inches. The second opening, *2*, is for the middle portion of the octagonal breech and, again, if a little deeper, could have served for the same position on a round barrel. However, at this point a difficulty arises; the opening *3* was for the end of the octagonal portion while the opening *4* was for the commencement of the round part of the barrel immediately after the octagonal. Nevertheless, the same basic principles apply and with regard to opening *4*, *by this the muzzle of the barrel shall be governed*, a factor which could only apply to a swamped barrel otherwise the barrel would have remained cylindrical from this point onwards. Thus, opening *5* applies to the reduced diameter *one couto* behind the muzzle.[6]

After turning it was polished with emery applied using blocks of hard walnut suitably 'hollowed'.[7] After polishing, the barrel *shall be taken to the drilling lathe, and be held between two tin clamps, so that it cannot be harmed, and with the short drills, which have cutting edges of two fingers or one inch, and there shall be opened in it the distance that the breech-plug is to have with thread, in such width* [diameter] *that after the thread is made in the barrel, it shall not be equal in height to the bore, so that when the cord is put inside, the thread is not an impediment to the lying of the cord, which has to lie evenly everywhere . . .*[8]

What is curious about this passage is that the barrel has already been bored and checked by 'the cord' so this reference to it seems out of place at this point.

The authors then go on to note that the drills shall be entered coaxially with the bore and once drilled the breech plug was entered coaxially. There is no mention of how the threads were formed but close examination of one illustration of the barrel maker's workshop in Fig. 4.1 shows what appear to be a 'screw-plates' [dies] along with two taps, Fig. 4.5.

Care was taken to ensure that the length of the threaded portion of the breech plug matched that in the barrel.

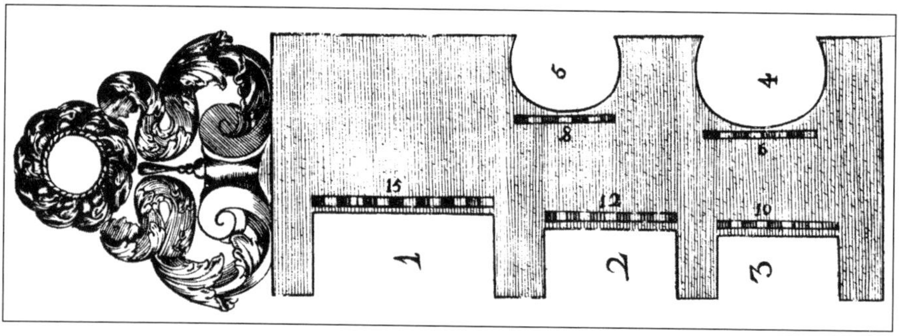

Fig. 4.4. Gauges for a barrel with bore of 14 *adarmes* (0.786 - 0.804 inches/19.9 - 21.3 mm), one *vara* (43 inches /1100 mm) in length and weighing approximately 3.5 *arreteis* (3.5 lbs. / 1.6 kg). (Fiosconi and Guserio, 1718, p. 84)

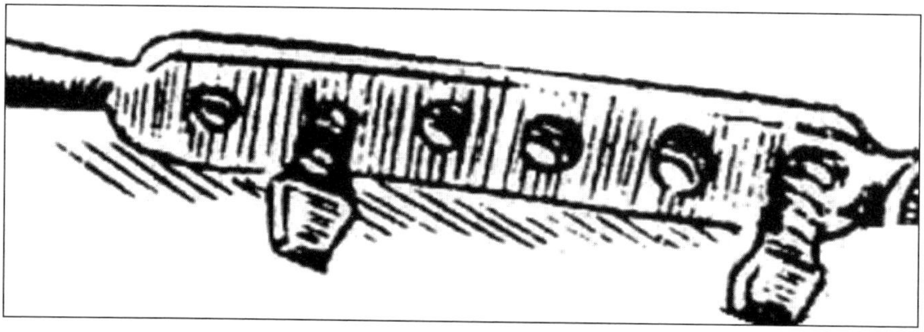

Fig. 4.5. An enlarged detail showing what appear to be a set of threading taps and a 'screw plate'. (Fiosconi and Guserio, 1718, p. 94)

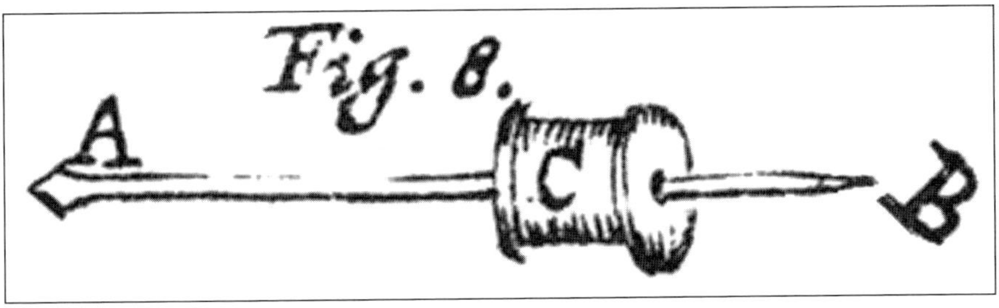

Fig. 4.6. A typical 'bow' drill shown by Moxon, the cutting end at A, a 'bobbin' C for the bowstring and the pivot, B, which fitted into a dimple on the breast-plate. These were used before, and even long after, the advent of the 'twist' drill, in many crafts.

When this stage had been reached, the barrel was then ready for test-firing, or 'proving' again. It would appear that this was carried out before drilling the vent and instead a special breech plug was used which could have a fuse fitted to ignite the charge and fire the gun, so that the actual breech plug was not damaged. Two charges of powder were loaded, each weighing the same as the bullets to be fired, followed by two bullets. Three shots were fired in this same manner and: *after the trial, everything must be as before, without inside or out there being seen the least scale or bristle, and if it is free of any defect, it will be good.*[9]

As noted earlier, it was 'double charged' and 'double shotted', showing this to have been a severe test.

Having finished the barrel and 'proved' it, the 'touch-hole' or vent was drilled. It was positioned by placing the breech plug on the outside of the barrel, matching its position when fitted, and a line scribed on the barrel adjacent to the face of the plug. Then at a distance of *the thickness of a testoon*[10] forward of this line, a centre punch was used to mark the position of the vent slightly below the centre line of the bore, *for the fire (which by nature rises) finding the touch hole to have been opened lower down, on entering will make the powder burn more rapidly.*[11]

The drill used also had the diameter the thickness of a 'testoon' and, although its nature is not specified, they were common tools and very much like the one illustrated by Moxon,[12] Fig. 4.6.

4.3 External finishing in the later 18th century - France and the M.1777 Musket

At the time *Espingarda* was being written, Villons in France in 1716 described a machine for draw-filing barrels in what appears to be a French patent.[13] As may be seen from Fig. 4.7, this was a water-powered machine, powering several 'filing benches' equipped with what Villons describes as a *milling cutter*. This cutter can be seen more clearly in a detail from this illustration shown in Fig. 4.8.

While Villons calls it a milling cutter, it is a rotary file since in the same publication that Fig. 4.8 appears, the author describes it as *the circular cutter on*

Fig. 4.7. A set of machines for external ‘milling’ of gun barrels (Villons, 1716).

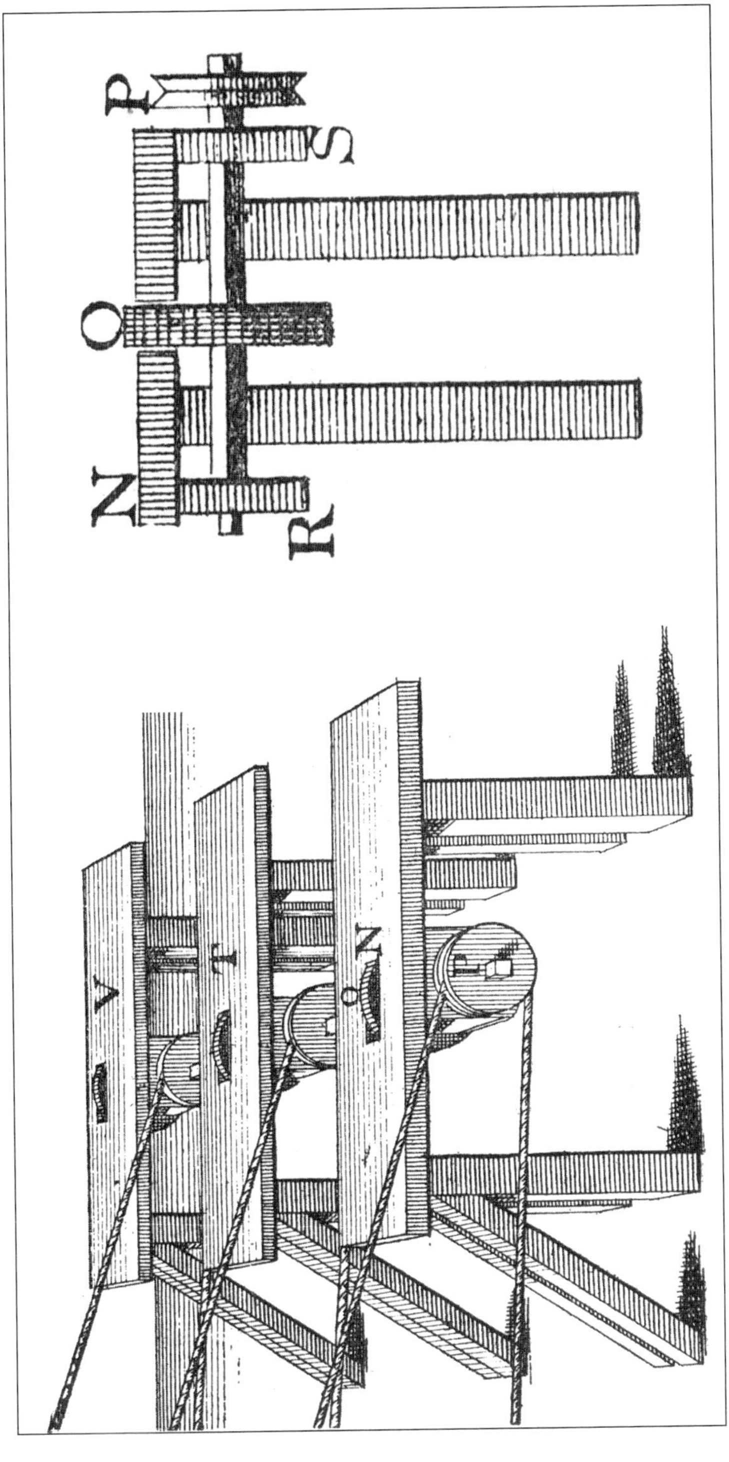

Fig. 4.8. Details of Villon's machine showing (right) the nature of the 'milling cutter'. (Taylor, 1920, p. 24)

which file teeth have been produced.[14] Two years later in 1718, the same year that *Espingarda Perfeyta* was published, Reamur described and illustrated a closely similar device which he referred to as a 'disc file' used for pointing pins. Such devices were still being made in the later 20th century and the author was able to watch a female file cutter in Sheffield making the 'milling' tools for key cutting machines.

While the Villons machine might embody the earliest reference to the use of rotary cutters in gunmaking, it is predated by Bion in 1709 who used a similar cutter in a wheel-cutting machine for watch making.[15]

Judging from the illustration and from the description, the machines of Villons were less sophisticated than those illustrated later by Diderot and appear to have had no adjustment for depth of cut. Also, it has to be presumed that with Villons' machine the barrels were simply held in the hand and moved over the rotating file.

From what can be gathered from the illustrations and brief descriptions accompanying them in Diderot's *L'Encyclopédie*, at least one new machine and an additional method had been introduced and which may have been used in concert. One, captioned as *machine a canneler les canons de fusil*, translates as 'machine for rifling gun barrels' but closer examination shows that the machine in Fig. 4.9 is unlike any rifling machine.

'*Canneler*' means to 'channel, groove, or flute'[16] which would correspond with rifling. However, it might conceivably relate to producing facets on the exterior. The most compelling evidence for it not being a rifling machine is that the barrel is held in its sliding carrier with centres inserted into the muzzle and breech of the barrel, Fig. 4.10, thus blanking them off and precluding the insertion of any rifling cutter!

An alternative view of the machine, Fig. 11, shows the guides in which the sliding carrier fits and the top edge of a rotary file or 'milling cutter', identical to that described by Villons, mounted below the barrel, and fitted with a pulley to be rotated by belt running on the large hand-operated driving wheel shown in Fig. 4.9.

Fig. 4.9. Diderot's illustration *machine a canneler les canons de fusil*. (Diderot, 1762, pl. II)

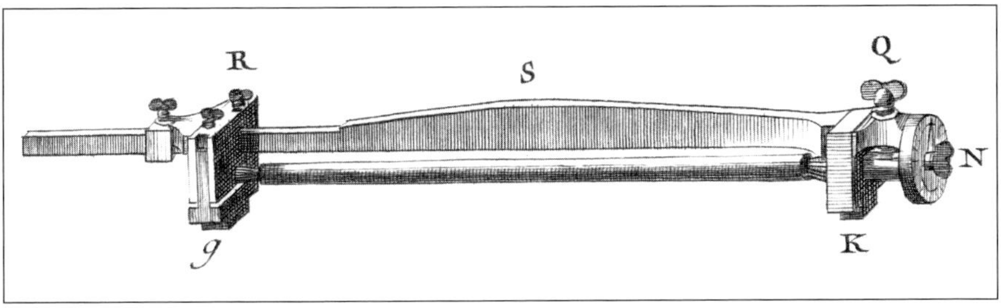

Fig. 4.10. Showing the barrel supported by centres inserted into breech and muzzle.

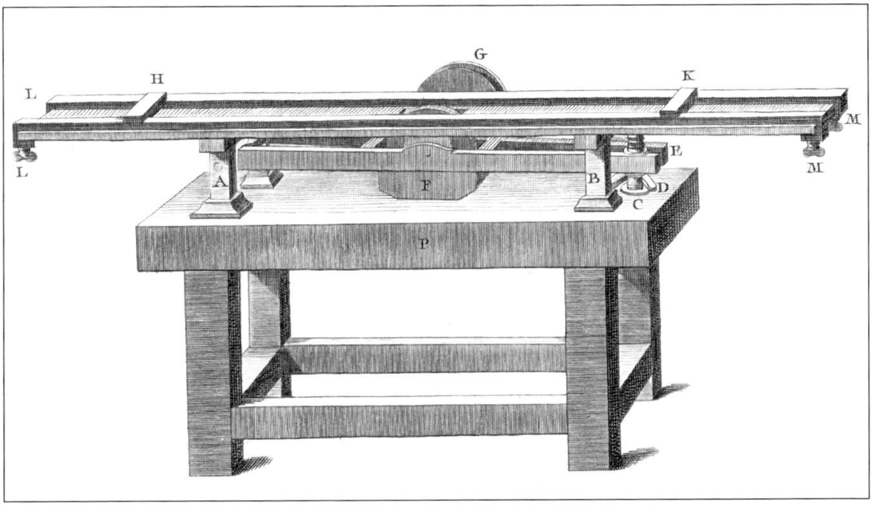

Fig. 4.11. View of the machine from the opposite side showing the carrier guides and the sub-frame on which the rotary file is mounted. (Diderot, 1762, pl. 2)

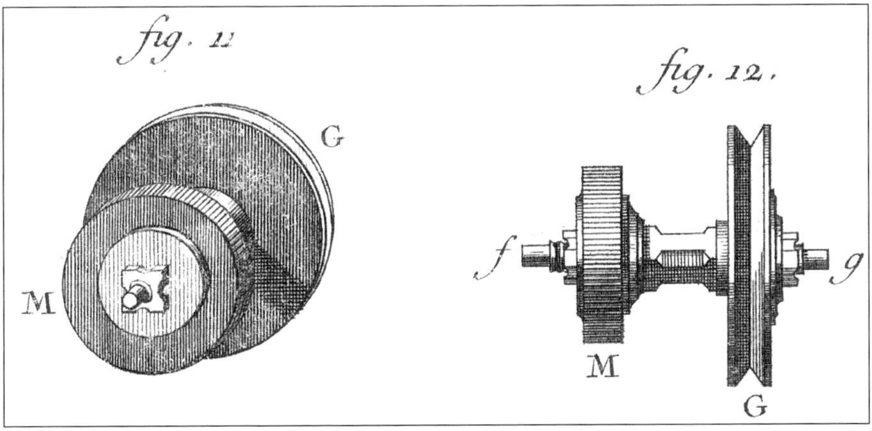

Fig. 4.12. Detail of the cutter on its arbour and fitted with a drive pulley.

This machine, however, shows an altogether higher degree of sophistication than that of Villons, possibly a result of experience in that machine's use. The cutter still operated on the underside of the barrel but was carried in a frame pivoted at one end on a support on the bench and with a screw adjustment for height at the other end, Fig. 4.13.

This feature allowed the 'depth of cut' to be easily adjusted. Also, the frame carrying the barrel allowed for adjustment to suit different barrel lengths and degree of taper.

It would seem that, when set up and the cutter put in motion, the barrel on its frame was slid along guides on the 'bench' by hand and over the cutter, as evidenced in Fig. 4.9.

Being a cylindrical cutter, it would cut flats on the barrel. None of the illustrations, as far as can be judged and as expected in a military musket, show

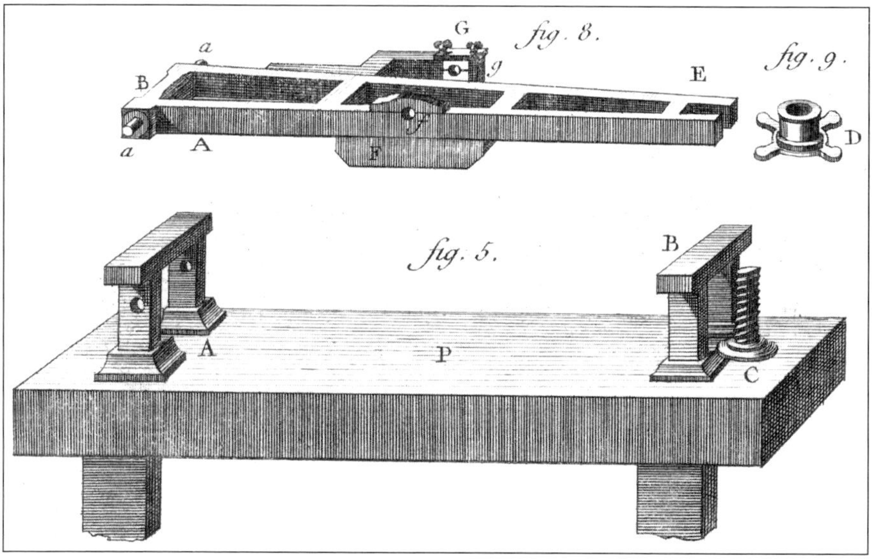

Fig. 4.13. The machine bench, showing the relationship between the 'cutter frame' and the bench with the height adjusting screw shown.

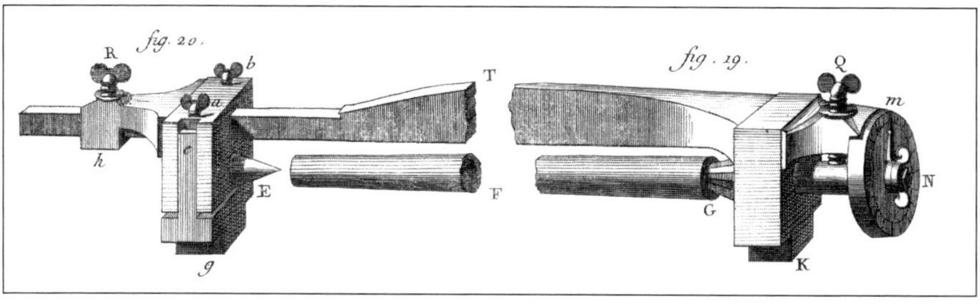

Fig. 4.14. Barrel support frame, with a barrel, F, mounted between centres, G and E, the centre E adjustable vertically and horizontally to allow for different barrel lengths and tapers.

anything other than round barrels blending in to octagonal facets at the breech. This machine was presumably used, therefore, to create a series of narrow facets on the round, tapered tube. At the right-hand end of the barrel frame can be seen what is taken to be a simple form of indexing device, allowing the barrel to be partially rotated in preparation for each passage over the cutter. Again, relying on what seems to be a reasonable assumption, rotating the barrel a small degree each time would create multiple facets along the barrel. Since a circle can be considered as a regular polygon with an infinite number of sides, the more of these facets created, the closer would the barrel approximate to a circular form – but exactly how many facets were produced is not known.

If it were possible to produce an infinite number of such facets, then the barrel would indeed be circular, but that is impossible in the real world. It would appear, then, that the second new process, grinding, might have been intended to take over at this point. It could be argued that it was an alternative process, but, if it produced satisfactory results, why bother with the filing machine at all? In some respects, it was a retrograde technological step after the care and thought that had been put into the machine just described which, with barrels on centres in the bore, would also lead to equality in wall thickness. Why a lathe was not used when such machines were obviously available is not commented upon. A detail from Diderot's illustration of the barrel boring workshop, Fig. 4.15, shows a barrel grinder at work in the top right corner. Unfortunately, whilst various parts are designated with key letters, no key accompanies the caption to the illustration! However, Cotty describes this highly skilled operation, in which the grinder maintained concentricity of bore and the external surface: *The barrel takes its external form from the sandstone wheel, which turns in a trough full of water that keeps it cool. The extension of the axis of this grindstone fits in the centre of a lantern gear from which it receives its movement . . . The grinder stands next to the grindstone, so that if it breaks, which takes place with extreme violence, he is less exposed. He begins to grind the barrel at the muzzle and continues to the end of the breech, where he regulates the pieces already formed in the forge. It is worth observing that the measures used by the grinder are a little above the exact dimensions, otherwise, the barrel would be put below size by polishing it.*[17]

In Diderot's illustration, Fig. 4.15, the barrel is shown on a mandrel with the outer end, S, resting on a rack with pegs, T. The opposite end is held by the grinder and would suggest that the barrel is held against the grindstone and allowed to rotate freely whilst being moved across the face of the stone so that the whole surface of the barrel is ground. How this was integrated with any filing process is not stated, but if the barrel had been previously filed, grinding would have removed any traces. Cotty then goes on to state: *The greatest attention must be paid to the fact that the wheels are very sound and, that they have no cracks; for if they have some defect, the high speed of rotation tends little by little to disunite the parts and they finally separate with the greatest velocity due to the effect of centrifugal force. The parts escape tangentially, taking away and breaking all in their path.*

The work of the grinder and that of the inspector is no less essential than that of the forger; these two, and especially the grinder, who reduces the diameter of

Fig. 4.15. Detail from the barrel boring workshop illustration showing a barrel grinder at work. (Diderot, 1779)

the barrel and which must be conducted with great care from breech to muzzle to produce a good barrel. It is therefore of the utmost importance that these workers are expert people.

The *inspector* referred to in the above statement relates to the last stage of this process: *When the barrel is brightened from one end to the other and it is well straightened with the cord, the inspector checks it with the compass for thickness and the calibre, and marks by check marks or file marks the places where the grinder has to remove some more material.*

The *compass* was *a steel rod folded in its centre, so that it forms two branches sufficiently close to each other. One of these branches is introduced into the barrel,*

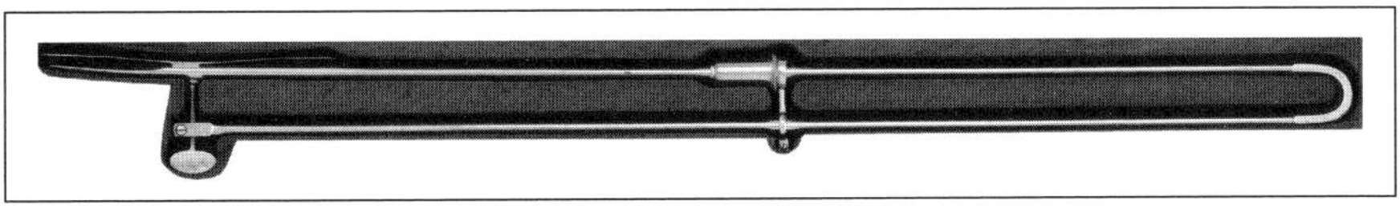

Fig. 4.16. Detail from Blanc's set of 1777 musket gauges, showing the 'compass' referred to by Cotty. (Photo (C) Paris - Musée de l'Armée, Dist. RMN-Grand Palais - © Pascal Segrette)

where it is held by means of a spring; the other branch goes down along the outside of the barrel and across its end is fitted a horizontal screw. By rotating the compass in the barrel, this screw indicates the places where iron must be removed. The length of the compass is a little more than half the length of the barrel.[18]

This *compass* might be the device seen in the top of Blanc's set of gauges for the M.1777 and illustrated in Fig. 4.16, which closely resembles that described in *Espingarda Perfeyta* as being used in Portugal in 1718.

Alder notes that attempts were made by Jean-Baptiste Javelle to develop machines for finishing gun barrels and reportedly patented in 1787.[19] One, '*de la machine à oucher*', scored the barrel at several points along its length to indicate the amount of metal to be removed by the barrel grinder. A second machine with a drawing titled '*la machine à aiguiser'*, is something of a mystery. The term '*aiguiser*', according to an English-French military technical dictionary,[20] applies to a grinding or sharpening process whereas in the drawing shown by Alder, and from his description, it is clearly a lathe and one with unusual features. The two main bed members are slightly angled so that any carriage holding a cutting tool moving along them would move inwards or outwards, enabling a taper to be produced on the workpiece centred between the two bed members. Unfortunately, the more intimate details of the machine are obscure and little more can be added. However, a machine with similar features will be encountered later when discussing a barrel tapering lathe patented in England by James and Jones.

A third machine in Alder's account is '*de la machine à faire les pans*', which was for forming the octagonal flats on the breech-end of the barrel. It was a simple milling machine with an indexing device to create regular octagonal facets.

Breeching

At this point, the barrel was almost finished. There remained the squaring of the muzzle and breech end, enlarging the bore at the breech and threading to receive the breech plug. Two items Diderot shows, but does not elaborate upon, are two breech plugs, Fig. 4.17; one is a forging and the other has the threads finished.

What is strange about the illustrations of 1779 and 1784 is that a left-hand thread is shown, which is unheard of. However, in a facsimile re-print (Diderot, Inter-livres) of an unspecified edition of Diderot's *L'Encyclopédie*, objects in illustrations, but not the captions, appear as a mirror image of the 1779 and 1784 versions. Thus, the thread on the breech plug correctly appears as a right-hand thread! The same engraver is acknowledged for all the illustrations and whose name is not mirrored! This discrepancy remains a mystery!

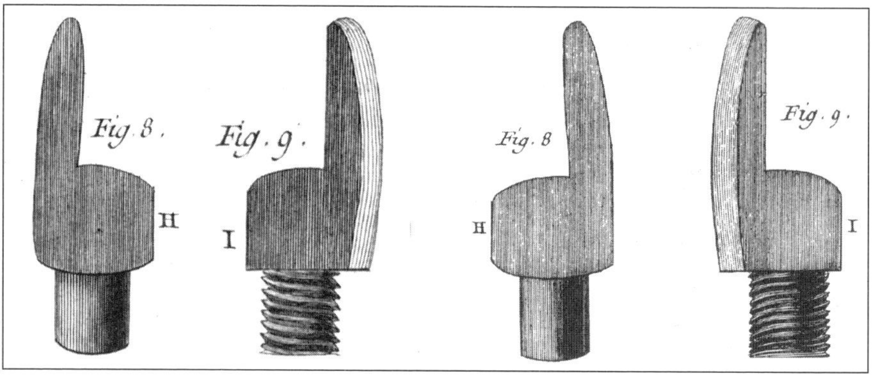

Fig 4.17. Re-orientated illustrations of breech plugs; *left,* from Diderot (1779 and 1784) show a left-hand thread: *right,* from a facsimile re-print (Diderot, Inter-livres) showing the correct right-hand thread.

The barrel, thus drilled and oiled, is returned to the '*garnisseur*', who cuts it squarely at both ends, forms the seat and thread for the breech plug and its tang and drills the vent. There is some difficulty in the interpretation of Cotty's term, '*garnisseur*' as to the exact role of this craftsman. In a French-English military dictionary[21] a '*garnisseur*' is simply described as 'finisher'. In the same dictionary, the term '*garnir*', a verb, is defined, amongst other things, as 'to stock'. On that basis, the '*garnisseur*' would seem equivalent to the '*monteur-équipeur*', or in English terminology, the 'setter-up'.

Once the breech-plug had been fitted, Cotty describes testing, or 'proving', in which *the barrel is tested horizontally, with two consecutive charges . . . on a bench, where it is placed so that the heel of the breech is against a strong piece of iron, etc. When this test is finished, it is removed, washed, dried, and treated externally with a smooth file and oil, until it presents to the eye end to end and to all the senses, only a very united surface.*[22]

One final test was required. The barrel was degreased and kept in a humid room for one month, after which it was inspected again to detect any flaws that might have been revealed and, if declared fit for service, was polished and inspected one last time before entering store. The testing of French military barrels was certainly rigorous.

4.4 External finishing in the early 19th century - Tula and the M.1808 Musket

From what we are told by Gamel, and in the absence of evidence to the contrary, Tula also seemed to have been at the forefront when it came to mechanising this aspect of barrel making, although some aspects were transplanted from England. Prior to Gamel's publication, it seems that the only method of external finishing of barrels was by grinding, in a similar manner as in France and, to judge from later practice, in England and America also. However, because of problems associated with grinding at Tula, it resulted in barrels with unequal wall thicknesses. Unfortunately, Gamel's narrative covering the external finishing process is rather confused. He notes that 20 lathes made by a Mr. Byrd of St. Petersburg in 1813,[23] were introduced for turning the outside of barrels. Initially, these lathes turned the barrel with taper along its full length but were subsequently modified to taper only the breech portion, the rest being cylindrical.[24] Why this form was adopted is not explained and is contrary to known practices – the musket in Royal Armouries' collection and a slightly shorter, and what might be termed 'light infantry' musket of 1828 in the author's collection, most certainly have barrels tapered along their full length.

To perform the operation, a mandrel was 'driven in' the barrel, followed by being set 'between centres' operated by screws [presumably only the tailstock centre], in the lathe.[25] The mandrel was rotated, and the driving of the barrel relied upon the friction between it and the tight-fitting mandrel. However, with this method the tailstock centre could be tightened to different degrees, causing the mandrel to bend slightly under the pressure exerted upon it leading to non-uniform wall thickness, *especially in the middle part. Therefore, on testing such barrels they exploded due to the unequal distribution of their strength. Machines of similar design, even in England, also had the same drawback in more or less degree.* Gamel also notes in a footnote that in England the best machines for turning barrels were made by John Lloyd for the government factory at Waltham Abbey.[26] From where Gamel obtained this information he does not say but Waltham Abbey was the government's gunpowder factory and the only barrels likely to have been encountered there would have been powder kegs!

The deficiencies of Byrd's machines were overcome by Jones' development of a machine in 1824, Fig. 4.18, which Gamel eulogises as being *considered perfect, since the possibility of getting a barrel with a non-uniform wall thickness is excluded.*[27]

Whilst superficially similar to having a boring bar under tension, placing a mandrel under tension made this machine conceptually very sophisticated. The barrel was carried, and free to rotate, on a steel mandrel, *3*, Fig. 4.19.

On the exterior of the breech-end of the barrel threads were cut square with the axis of the bore, and on this was screwed a special collar, *2*, Figs. 4.20 and 4.21, having ratchet teeth on its rear face. These teeth engaged with matching teeth on an extension of a gear carried on a shaft which formed the 'headstock' spindle. The steel mandrel had a tenon formed on the breech-end which engaged with a slot in the spindle, *10*, and was secured by a pin, *8*.

At the muzzle-end of the barrel, Fig. 4.21, the mandrel, *3*, was fitted with a sleeve, *4*, having the same external diameter as the finished muzzle.

The muzzle-end of the barrel and mandrel assembly was passed through the triangular opening in the 'tailstock', *6*, which was a vertical extension of the lathe mounting like the headstock, and the tenon on the breech-end fitted into the mortice, *10*, on the spindle nose.

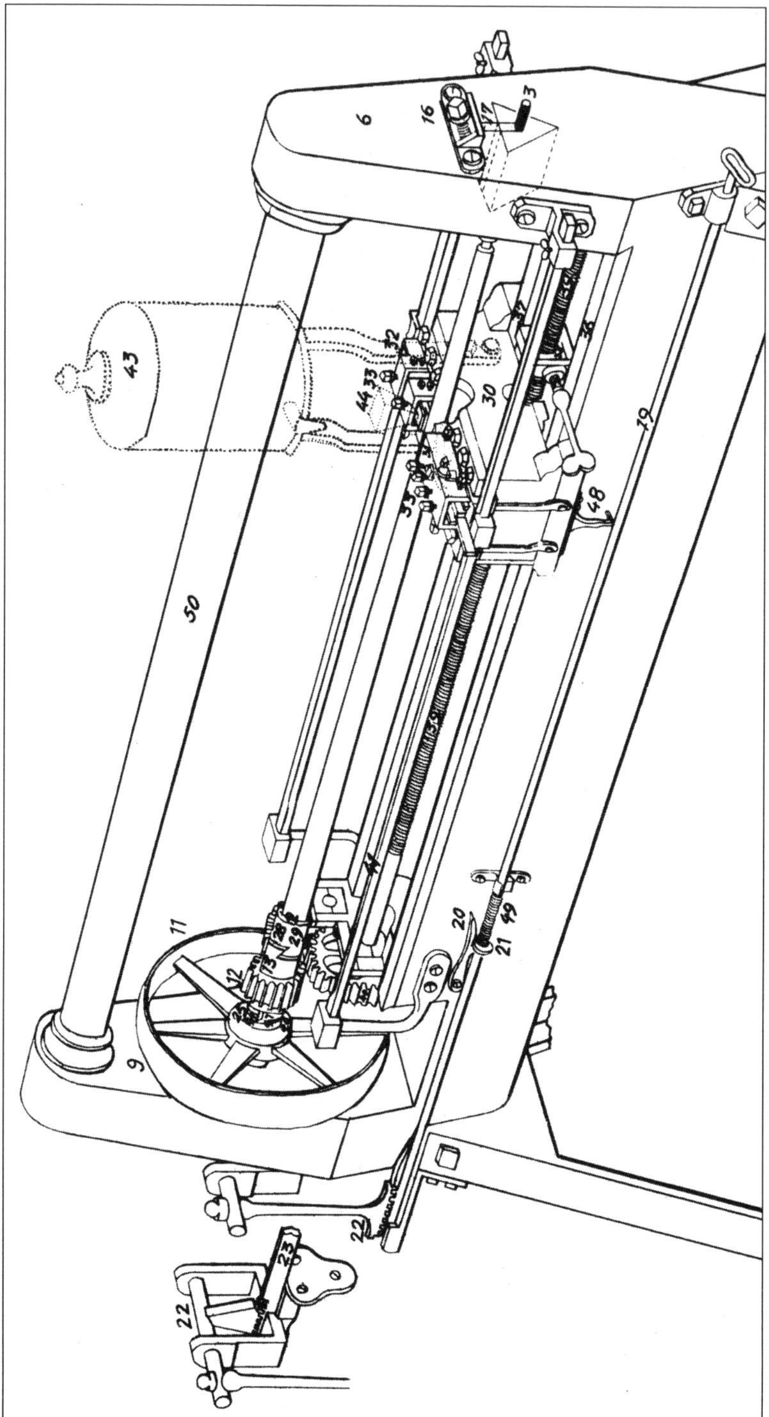

Fig. 4.18. Jones' lathe for turning barrels. (Gamel, 1826, fig. VII)

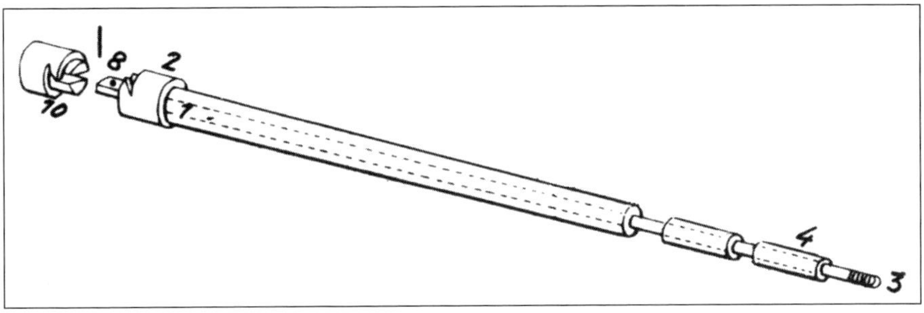

Fig. 4.19. The barrel on its mandrel. (Gamel, 1826, fig. VII)

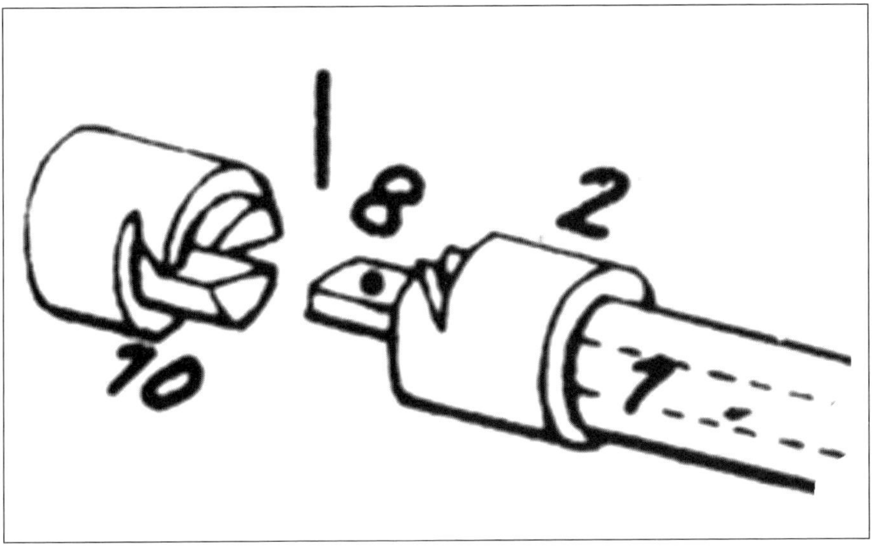

Fig. 4.20. Detail from Fig. 4.19 showing the breech end of the barrel, mandrel and the spindle nose.

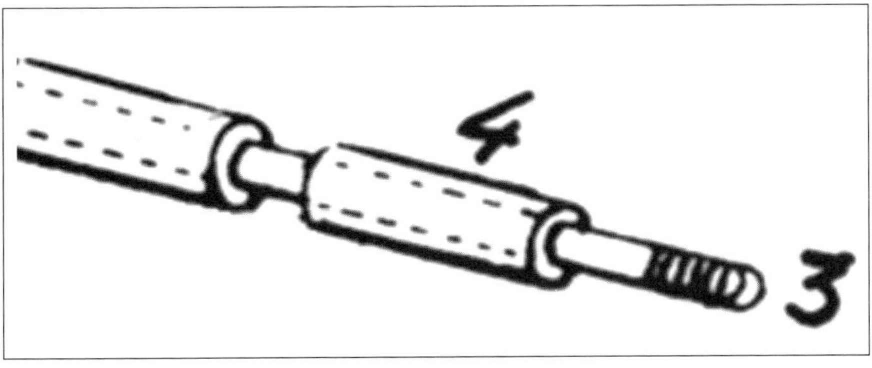

Fig. 4.21. The muzzle-end of the barrel and mandrel.

A prismatic block, Fig. 4.22, was then slid onto the mandrel and into its triangular housing in the 'tailstock', at which point the securing pin, *8*, at the breech end could be safely fitted.

It will be noticed that the hole in the prismatic block is placed eccentrically, and the block is offset in its housing to maintain the centreline of the barrel. This eccentric positioning of the mandrel allows the flanges of the nut, *14*, to overlap the two edges of the block, Fig. 4.23, and to bear upon the faces of the upright 'tailstock' body, putting the mandrel under tension against its retaining pin, *8*, at the opposite end.

A steel wedge, *18*, was pressed into the slot in the top of the prismatic block, Fig. 4.24, by a screw, *16*, acting through an intermediary block, *17*, so that the

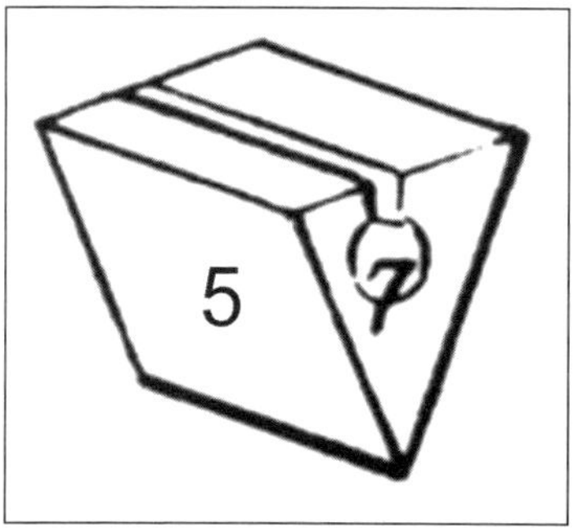

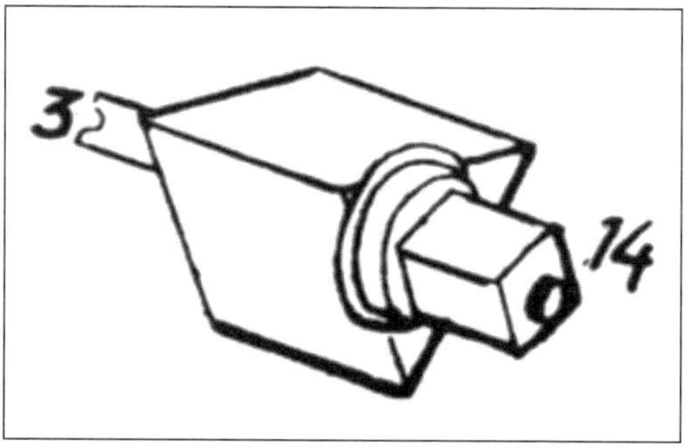

Above left: Fig. 4.22. The prismatic block carrying the outer end of the mandrel. (Note that in Gamel's original drawing this was mistakenly numbered as 3.)

Above right: Fig. 4.23. The prismatic block (slot omitted in original drawing) with the flanges of the nut, 14, overlapping two edges.

Fig. 4.24. The system of wedging the block in its housing in the tailstock.

pressure from the screw was transmitted downwards on the wedge to expand and tighten the closely fitting block in its housing, providing additional rigidity to the assembly.

When ready, a simple mechanism, Fig. 4.25, was used to slide the belt pulley and the pinion, *12*, with its 'dog-clutch', *13*, forward into engagement with the dog-clutch on the barrel.

This also engaged the gear on the central leadscrew, and half-nuts on the carriage allowed the leadscrew to be engaged at will. The carriage, Fig. 4.26, carried a turning tool and a steady, which were acted upon by their respective guide bars set in front of and behind the barrel.

The comment that *On this excellent machine, twenty barrels are machined per day,* and a footnote on the same page noting *The chips cut by the tool from the barrel, frequently forming an extremely long spiral thread*[28] indicates a high-quality in both the iron for the barrel, the cutting tool, and in the machine which could produce such very fine turning.

It automatically disengaged the carriage feed by means of a projection, *48*, Fig. 4.27, on the carriage coming into contact with a latch, *20*, and lifting it, allowing the spring, *49*, to push the rod to the left and disengage the drive by acting on the lever, *22*, shown in Fig. 4.25.

This is a more sophisticated machine than that of Nash [see later] for two important reasons: it utilised a system that enabled both cutting tool and steady to automatically create and adjust to a tapered profile; it maintained the cutting tool and steady on the centre height of the barrel.

It was at this point that the stud/tenon on the muzzle for the bayonet was fitted. The barrel was fitted horizontally in a vice and, using a reference gauge, two marks were made and the space between them filed to create a 'dovetail' slot. The bayonet

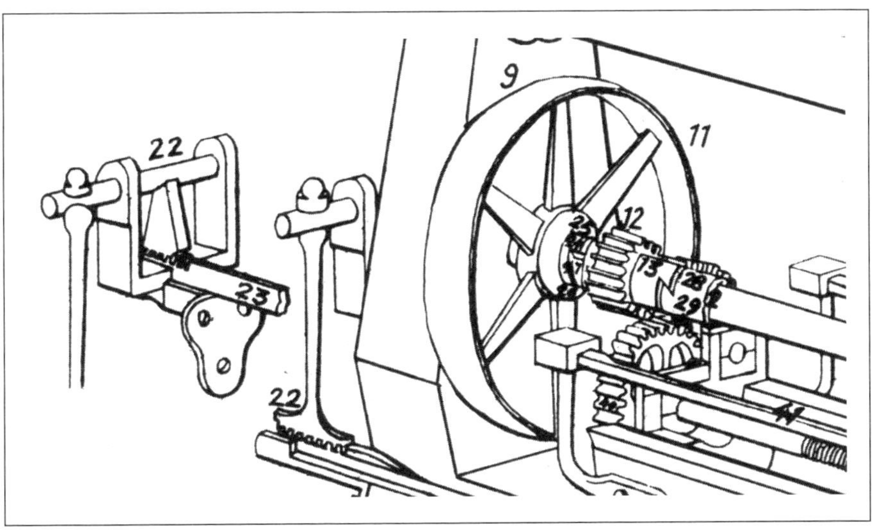

Fig. 4.25. Detail of the headstock and drive.

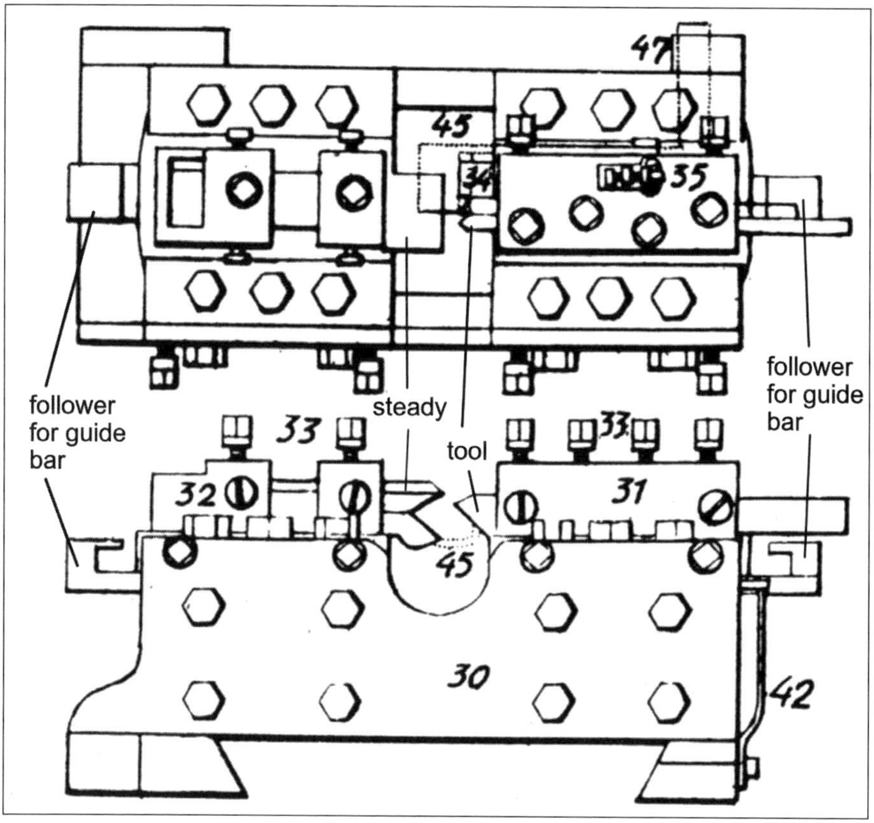

Fig. 4.26. Plan and elevation of carriage, showing the sliding holders for cutter, 31, and the steady, 32, which are acted upon by the guide bars to cause both to follow a pre-set profile.

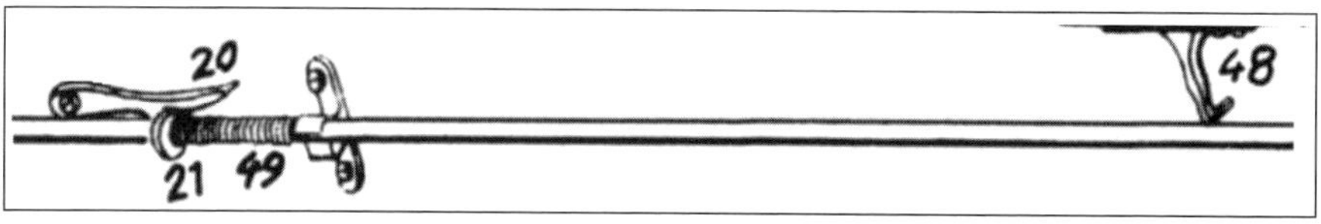

Fig. 4.27. Auto-disengage mechanism; when the projection, 48, contacted the latch, 20, it allowed the spring, 49, to move the rod to the left and disengage the drive.

stud/tenon, with matching dovetail and slightly over length, was pushed into this slot. The muzzle was then filled with clay, the lug covered with a mixture of brass chips or filings and borax flux, then placed in the hearth to achieve the brazing.[29]

Breeching

The breech had a thread of eight and one half turns[30] and threading the barrel and making the breech plug involved three innovative machines designed in 1826.[31] The barrel breech was first prepared by 'facing' which was carried out by the machine in Fig. 4.28.

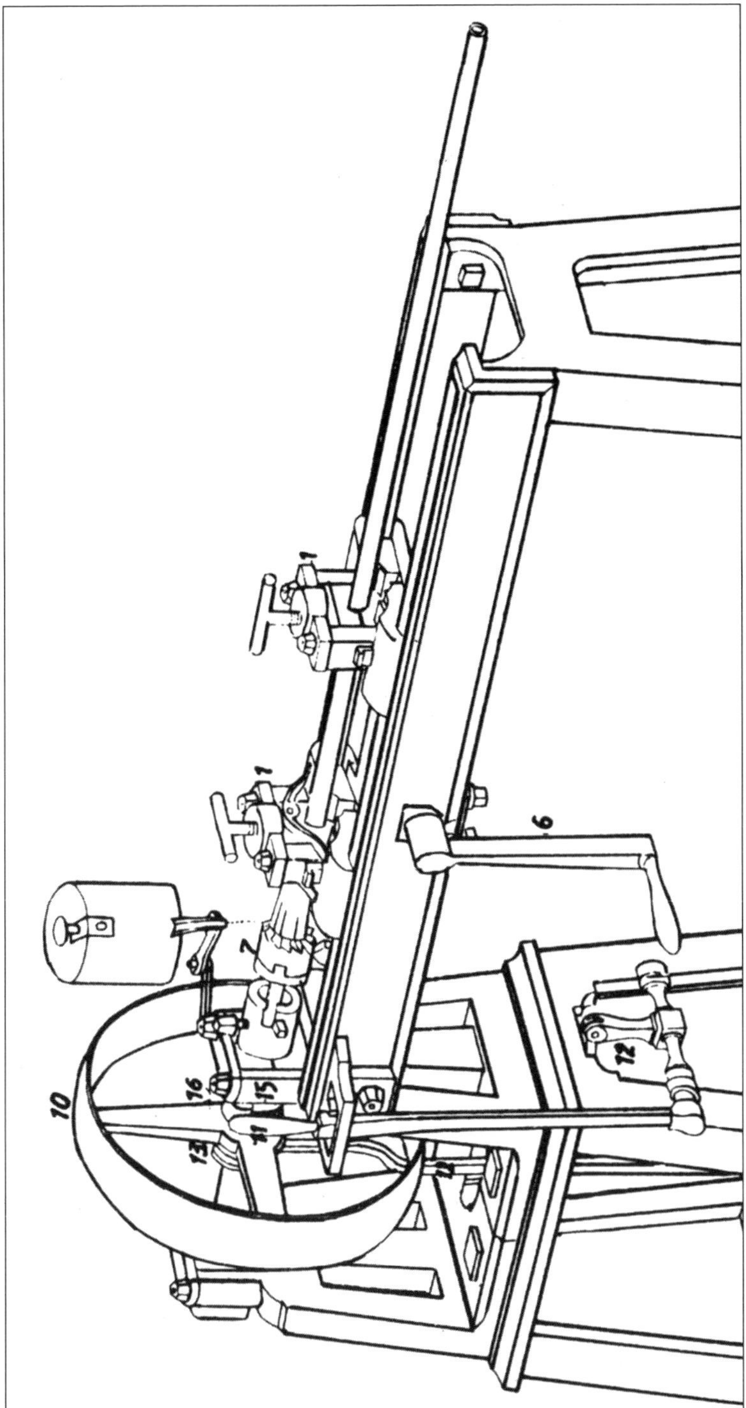

Fig. 4.28. Breech facing and drilling machine. (Gamel, 1826, Fig. IX)

The barrel was clamped in a sliding carriage fitted with two clamps, *1*, *1*, for securing the barrel and which could be moved by a handle, *6*, which carried a pinion engaging with a rack, *3*, beneath the carriage, Fig. 4.29.

It is presumed that the clamps were such that the barrel was clamped with its bore centreline corresponding with the axis of the machine spindle. The headstock spindle, *15*, carried the large driving pulley, *10*, equipped with a sliding 'dog-clutch', *14*, so the drive to the spindle could be easily disengaged. This spindle nose was fitted with a socket, *9*, having a slot to engage with a drive peg on the cutter arbour, *8*. The cutter, *7*, was effectively a face-mill and simply slid onto the spigot of the arbour and was driven by two raised keys on the flange of the arbour.

The protruding spigot of *8* acted as a pilot to align the face of the cutter square with the bore and when the machine was set in motion, the barrel was fed onto the revolving cutter by means of the hand-crank handle, *6*, operating through the rack and pinion. Once the breech had been faced, it was then counterbored on a similar machine, or on the same machine by simply replacing the facing cutter with a counterbore having an integral pilot, *18*, and adjustable depth-stop, *19*, Fig. 4.32.

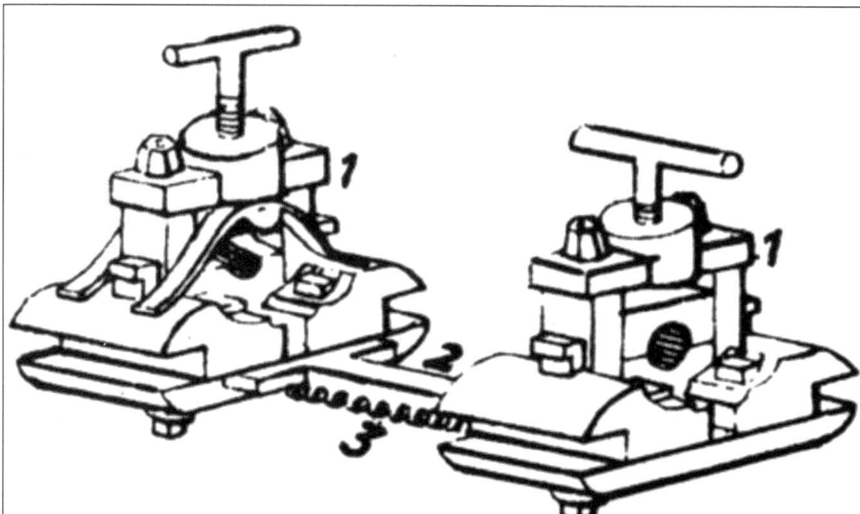

Fig. 4.29. The carriage on which the barrel was fixed for facing the breech. (Gamel, 1826, Fig. IX)

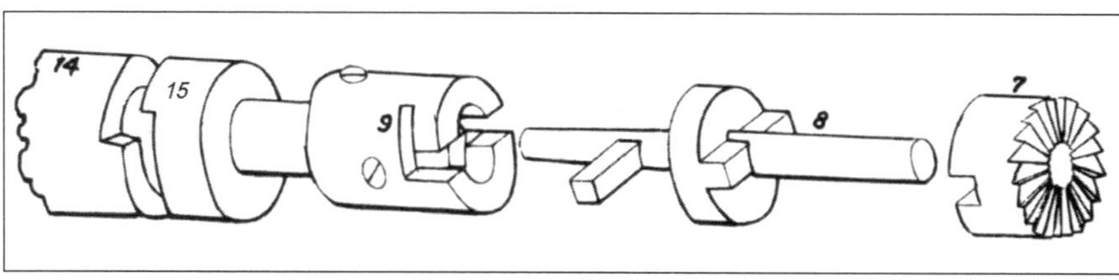

Fig. 4.30. The machine spindle and cutter assembly. (Gamel, 1826, Fig. IX)

Fig. 4.31. A breech-facing cutter of the type shown as item 7 in Fig. 4.30, probably English 19th century. (Author's collection)

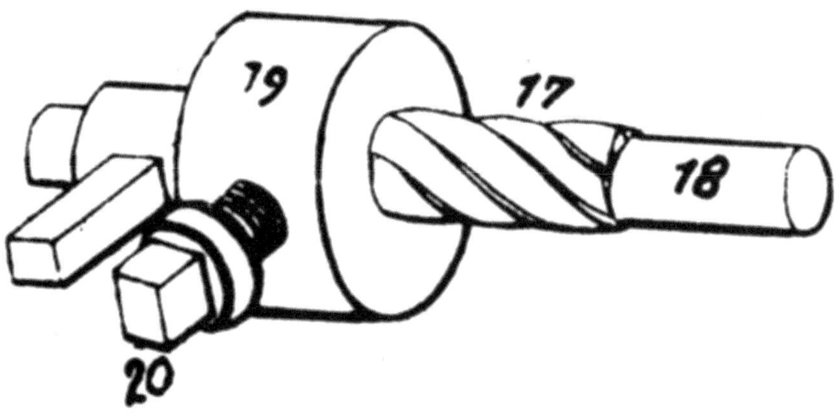

Fig. 4.32. Counterbore tool. (Gamel, 1826, Fig. IX)

An English device of similar nature and probably later date is shown in Fig. 4.33. It differed principally in having a rectangular 'tang' which would have fitted into a slot, rather than a circular tang fitting into a round hole, and a prolonged 'pilot'. This may have allowed it to be in some degree 'self-aligning' with the tang capable of sliding in a slot as opposed to being confined in a round hole. The facing and counterboring operation each produced two-hundred and forty barrels per day.

The breech was then ready for tapping, which was performed on another machine developed in 1826, Fig. 4.34.

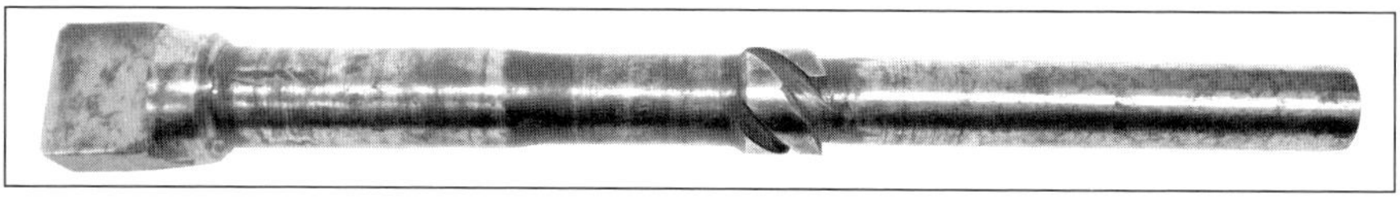

Fig. 4.33. Breeching counterbore, English, probably mid-late 19th century. (Author's collection)

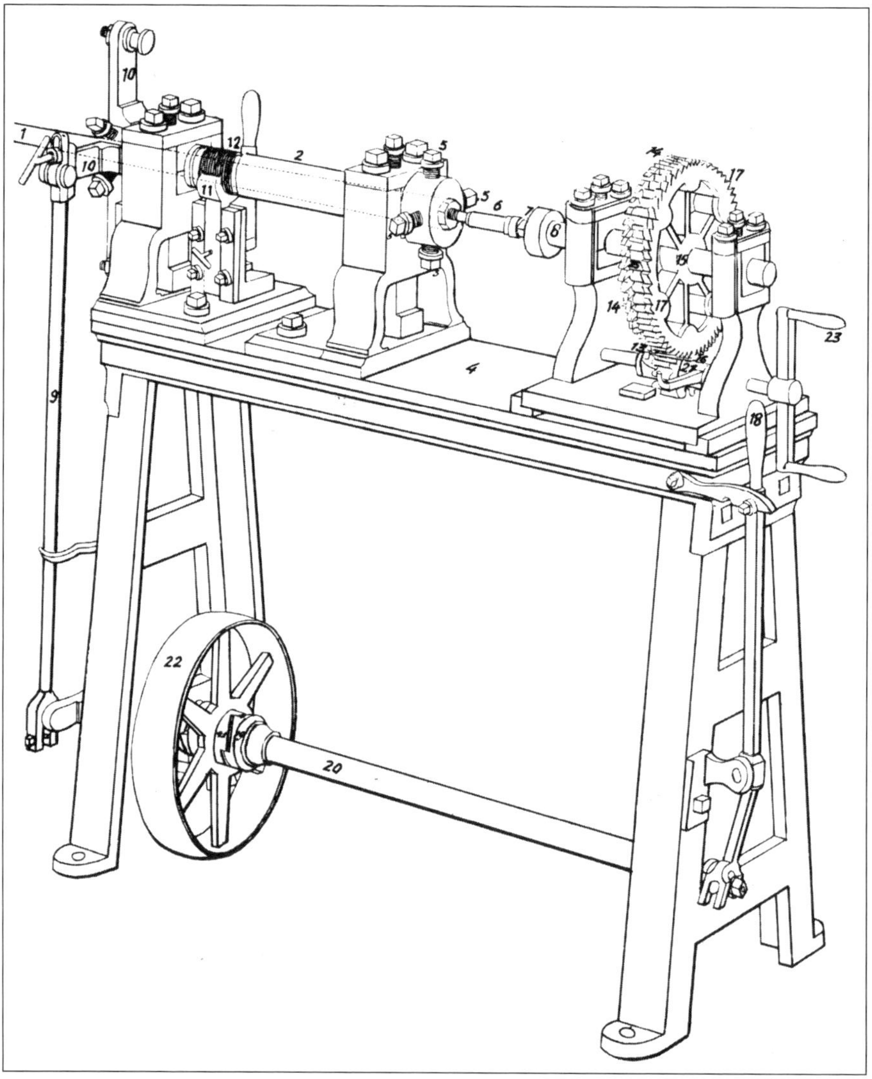

Fig. 4.34. The breech tapping machine. (Gamel, 1826, Fig. X)

The headstock was equipped with a large diameter, hollow spindle, *2*, into which the barrel was inserted, and its breech-end centred and held by means of screws, *5*, in a rudimentary 4-jaw chuck, often referred to as a 'bell-chuck', on the spindle nose. Centering was aided by means of a stub-mandrel, *3*, mounted on a compound slide, Fig. 4.36, fitted to the bed at *4*, so that its centre line corresponded with that of the headstock spindle.

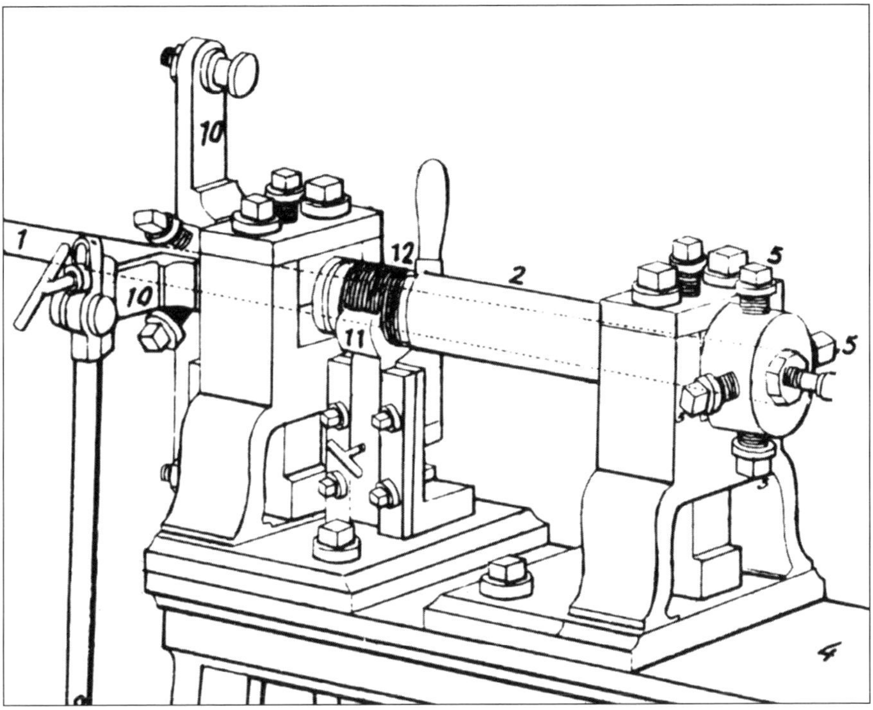

Fig. 4.35. The headstock of the tapping machine.

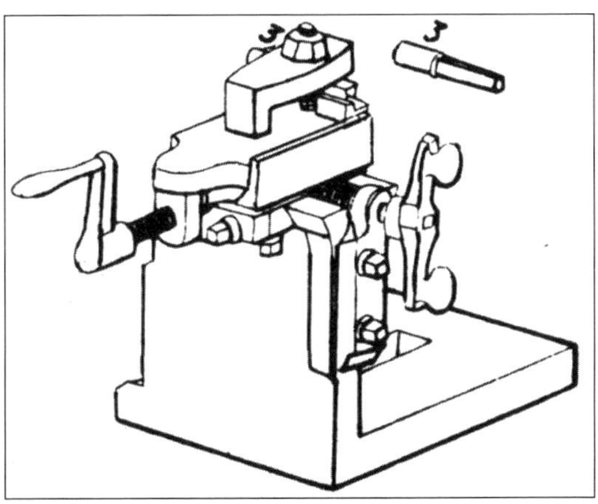

Fig. 4.36. Compound-slide with a stub-mandrel, 3, for centering the barrel.

Having secured the barrel, the tap, *6*, was fitted into the square socket on the 'tailstock' nose, *8*, Fig. 4.37. The tailstock barrel carried a gearwheel, *15*, flanked by two opposing ratchet wheels, *14* and *17*, each with their own pawl.

An English example of a tap for performing the same job is shown below, Fig. 4.38. Again, it has a rectangular tang but the thread it cut would be tapered at the end opposite the breech opening and may have been followed by a tap which finished the thread to full depth. It also differs from modern taps in have an obtuse cutting angle which cannot have been very efficient and possibly acted more in a 'thread rolling' fashion!

The headstock spindle, Fig. 4.39, was fitted at its outer end with a four-armed 'crank', *10*, and set screws in its hub were used to centre and support the barrel, *1*.

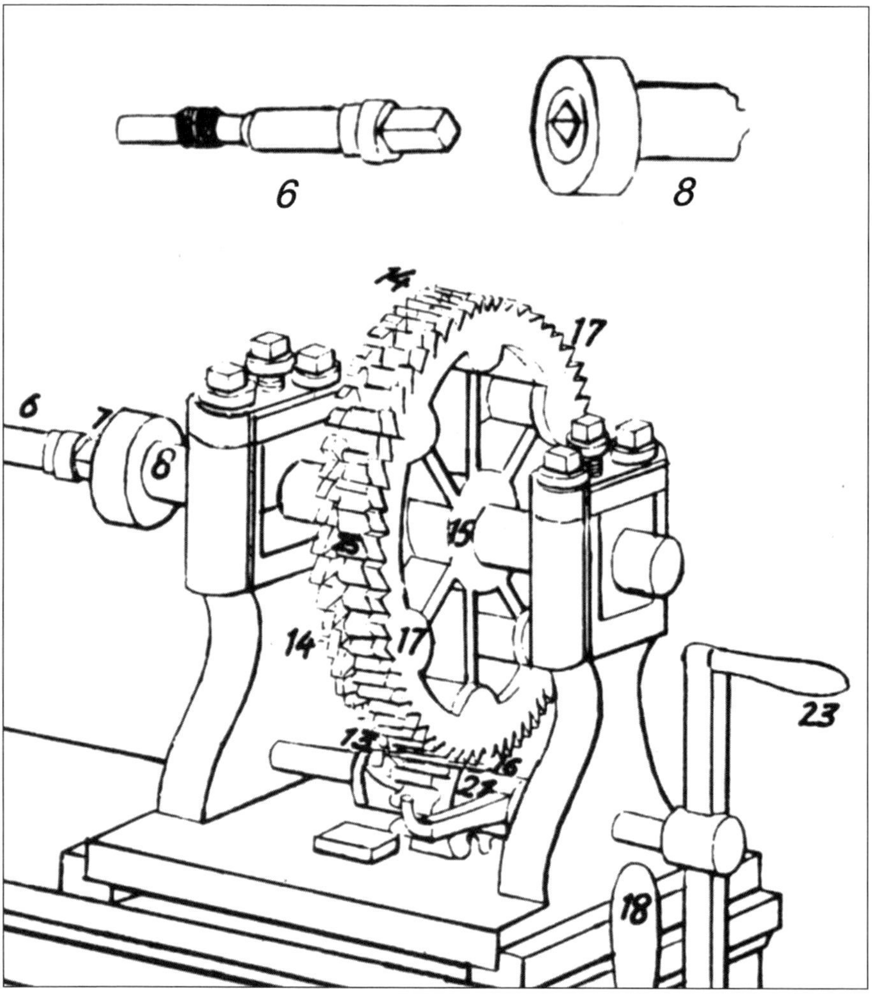

Fig. 4.37. The 'tailstock' with the tap, 6, and spindle nose, 8. (Gamel, 1826, Fig. X)

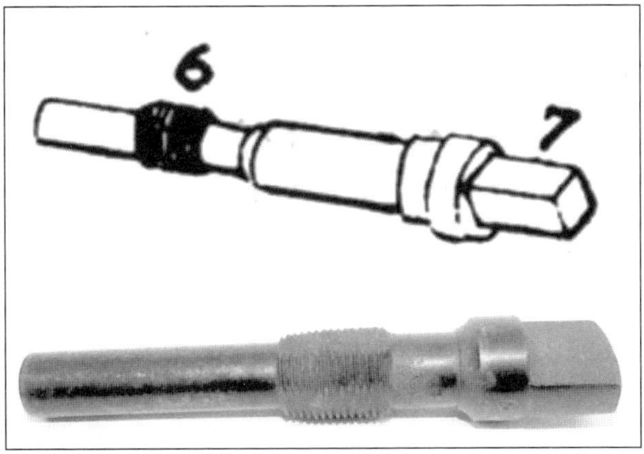

Right: Fig. 4.38. Breeching tap, English, probably mid-late 19th century. (Author's collection)

Below: Fig. 4.39. Detail of the outer end of the headstock showing the threads, 12, on the spindle and the half-nut, 11, the connecting rod, 9, and the four-armed crank, 10.

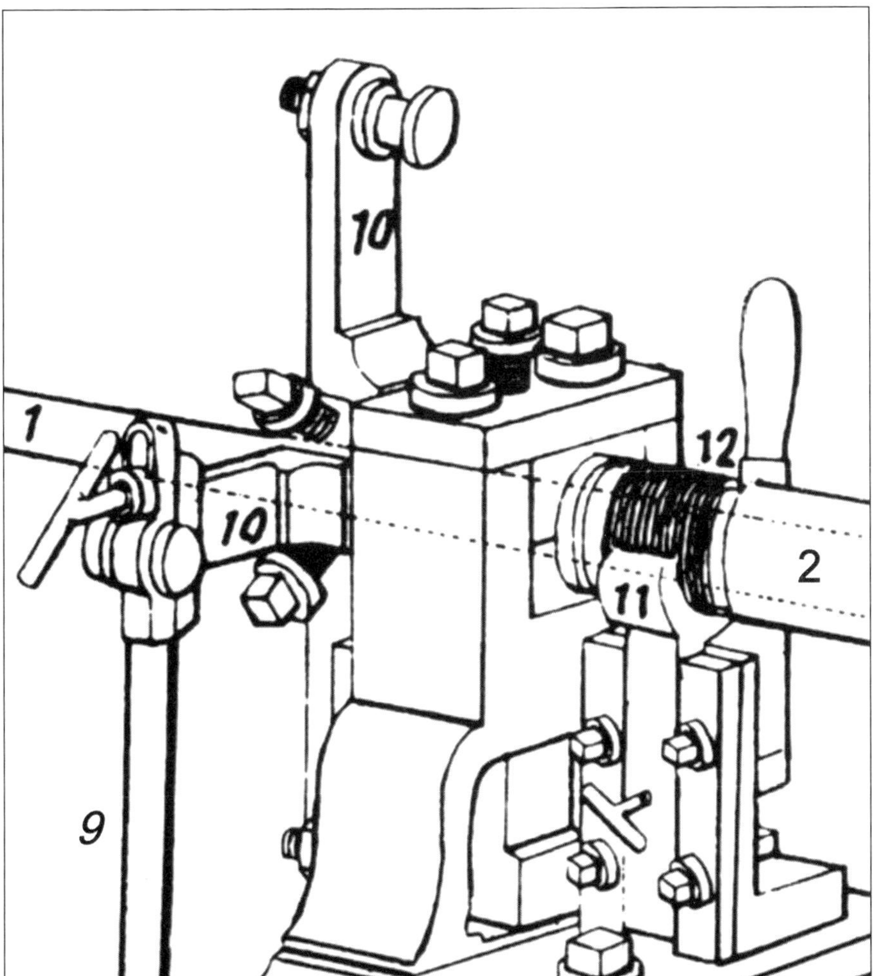

One of the arms was linked to a connecting rod, *9*, which reciprocated and gave the spindle approximately 1/6 turn in each direction, but how this reciprocating motion was created is not stated. It has to be assumed that it is connected with some form of crank on the drive wheel shaft positioned below the bed, like the machine in Fig. 4.34. The headstock spindle also has a thread, *12*, the same pitch as the breech plug tap machined on it and could be engaged with a half-nut, *11*.

The 'pilot' of the tap was inserted into the breech and the spindle and barrel set in its 'reversed' position, i.e., with the front leg of the crank in its uppermost position. Setting the machine in motion then caused the tap, its forward rotation being prevented by the ratchet on the tailstock barrel, to begin threading itself into the barrel. On the rearward stroke of the connecting rod, the headstock spindle would be rotated backwards, and the nature of the ratchet on the tailstock, which prevented forward rotation, would allow the spindle to carry the tap with it on its backward rotation. The next forward rotation of the headstock spindle would draw the tap deeper into the breech, and thus the process was continued until the correct length of thread had been cut.

The tap was withdrawn, either by disengaging the tailstock ratchet which prevented forward rotation and engaging the ratchet which prevents rearward rotation and putting the machine in motion; or it could be done faster by manually disengaging both ratchets and using the handle, *23*, to rotate the tailstock barrel via its gearing, as in Fig. 4.37.

It is instructive to examine the forging of the breech plug, Fig. 4.40, since it demonstrates many of the techniques used in forging generally.

The techniques and tooling used were more of Jones' innovations (Gamel, 1826, p. 112) and probably reflected English practice. The forging was performed using an anvil which incorporated various features, in conjunction with a variety of hand-tools. The different stages are illustrated below.

The first step was to take the iron bar and insert it into a simple gauge mounted on the anvil to mark the upper edge with a chisel, indicating the length the 'butt' was to have. It was then heated, and the 'butt' forged between two semi-cylindrical swages, one being fixed on the anvil's top face. The piece was rotated during this process to give it the correct shape. It was then heated again, and the bar slid down a fork, leaving the butt projecting which was then hammered on its face to form

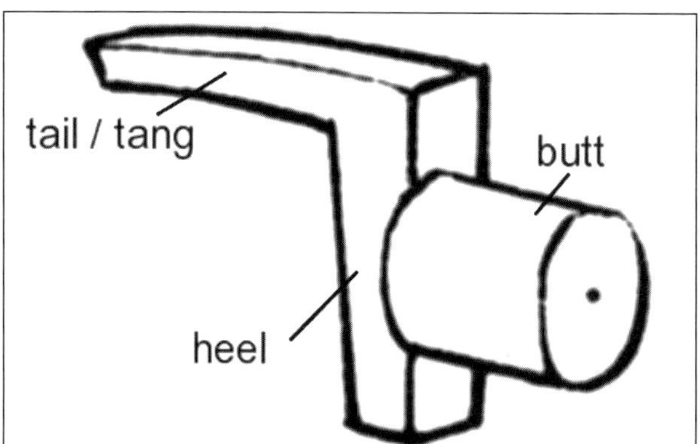

Fig. 4.40. The Tula breech plug and its terminology. (Gamel, 1826, Fig. XII)

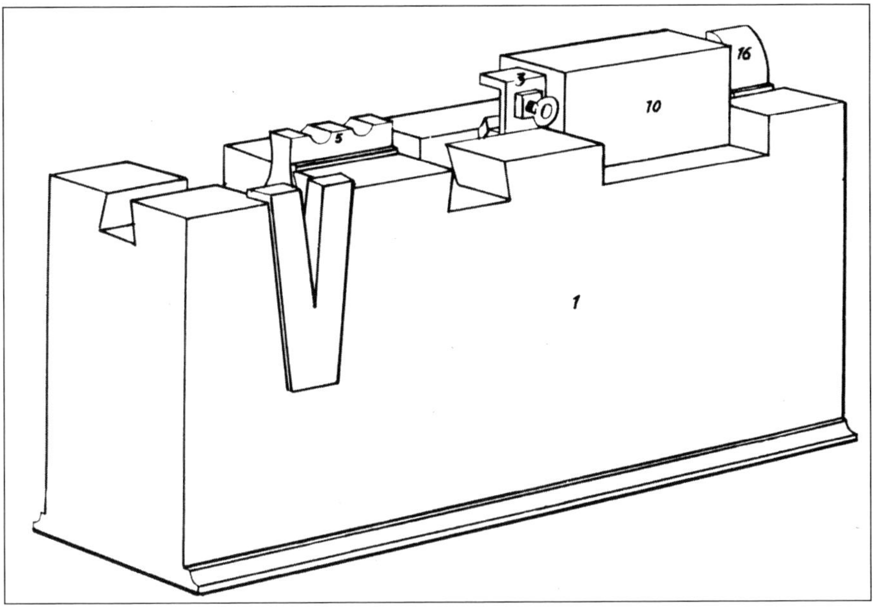

Fig. 4.41. Jones' form of anvil introduced at Tula. (Gamel, 1826, Fig. XI)

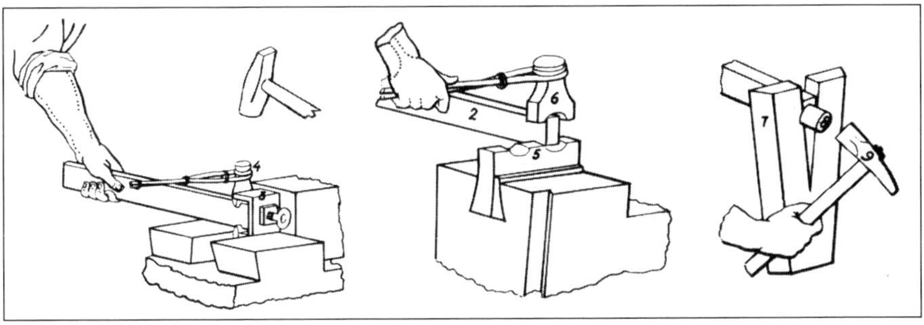

Fig. 4.42. *Left*: marking the length of the 'butt'; *Centre*: swaging to form the cylindrical butt; *Right*: squaring the shoulders of the butt. (Gamel, 1826, fig. XI)

square shoulders at its junction with the forks of the bar. The bar was again heated, and a notch made to indicate where it would be divided to form the heel, leaving sufficient to form the tail, or tang. While hot, it was then separated from the bar with a chisel, grasped in tongs and the heel portion straightened on the anvil face using a flat fuller. Gamel speaks of this process being carried out in reverse order[32] but the illustrations clearly show the piece has been cut from the bar before straightening. The stub of bar then needed to be divided to form the tang and heel, Fig, 4.43.

The heated metal was placed with the notch made earlier on the edge of an upright chisel set on the anvil and dividing until the correct depth had been cut, followed by bending the tail or tang portion to approximately 90°. The butt was then placed in a die and the heel formed.

In finishing the breech plug, it passed through five stages; the filing of the heel; the marking of the centres of the butt and at a corresponding position on the rear face of the heel; the turning and the threading of the butt. Gamel states that at the time of his writing, a machine for filing the sides of the heel was under construction but offers little detail. Another small machine was used for making 'centre' dimples in the butt and heel, but again no further details are offered. When it came to the turning and threading, two machines were used. The machine for turning the breech plug is shown in Figs. 4.44 and 4.45 and is basically a specialised lathe.

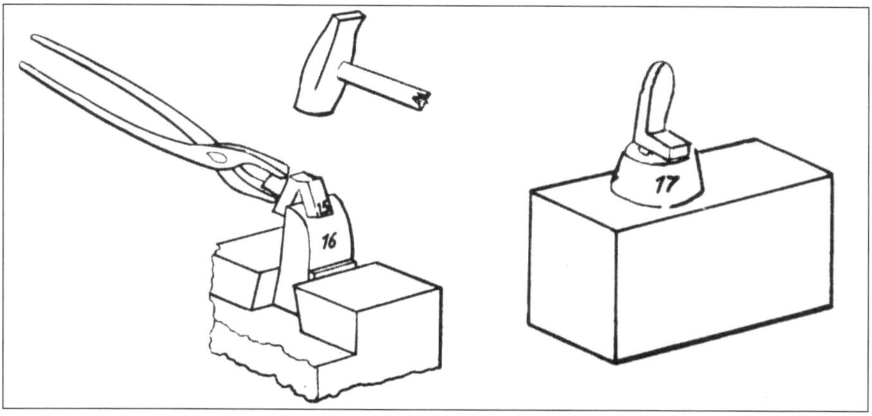

Fig. 4.43. *Left*: dividing the bar behind the butt to form the tang; *Right*: forming the heel.

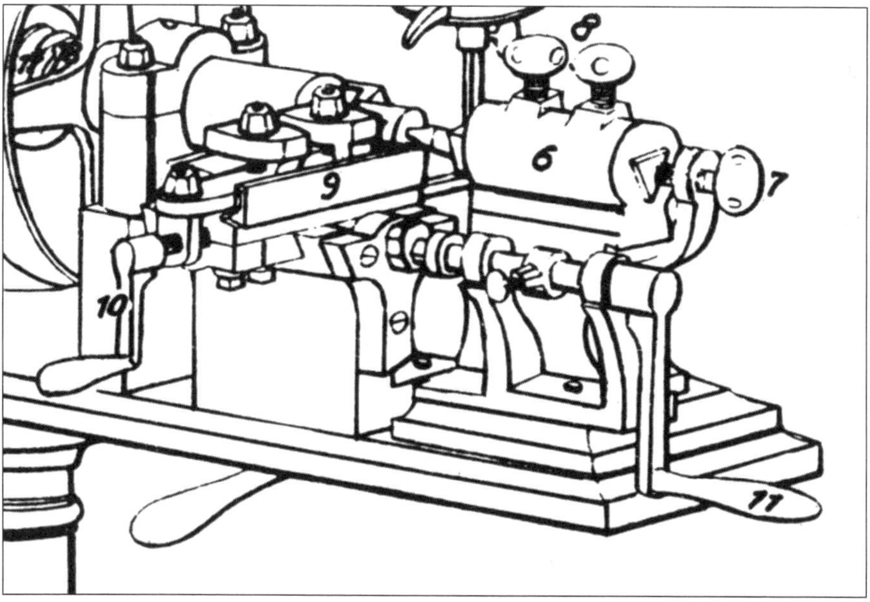

Fig. 4.44. Detail of the butt-turning machine.

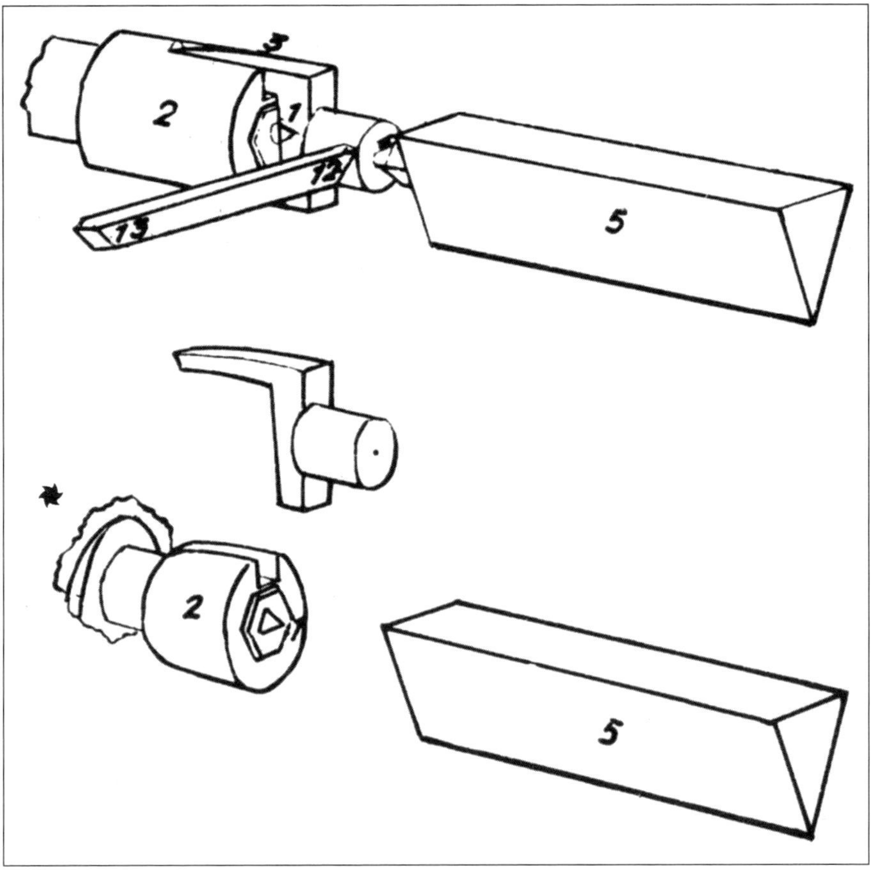

Fig. 4.45. Elements which fix the breech screw in position, and a turning tool, 13. (Gamel, 1826, Fig. XII)

The headstock carried a spindle whose nose was equipped with a holder to accept the tail/tang and heel of the breech plug and was fitted with a centre to engage the dimple in the centre of the heel for correct alignment. The tailstock barrel was a triangular prismatic bar, *5*, having at its tip another centre point which engaged the dimple on the butt of the breech screw by means of thumbscrew, *7*. Once fitted, the bar was clamped in position by thumbscrews, *8*. The tail of the plug fitted in a slot, *3*, on the spindle nose, *2*, and acted as a drive-dog in transmitting motion to the breech plug. The arrangement should be apparent from Figs. 4.44 and 4.45.

The device shown as *9* in Fig. 4.44 functioned as a compound slide carrying a turning tool, 13 in Fig. 4.45, and was used first to turn the butt cylindrical and then to face the end and the exposed faces of the heel forming the shoulders each side of the butt, its length and diameter being checked by gauges. The machine could turn 100 breech plugs per day.

The breech plugs were then threaded using a second machine, Fig. 4.46.

As in the previous machine, the mandrel nose was such that the breech plug could be fitted into a slot, *1*, the 'butt' tight against the end face, *2*, and held in place by a sliding collar, *3*, Fig. 4.47.

A 'chaser thread', *5*, was cut on the left of the mandrel and a set of dies, *4*, *4*, in the 'tailstock', both having the same thread as required on the breech plug. Fig. 4.48, during operation, the 'chaser thread', caused the mandrel to traverse to the right, causing the butt of the breech plug to be positively fed into the dies.

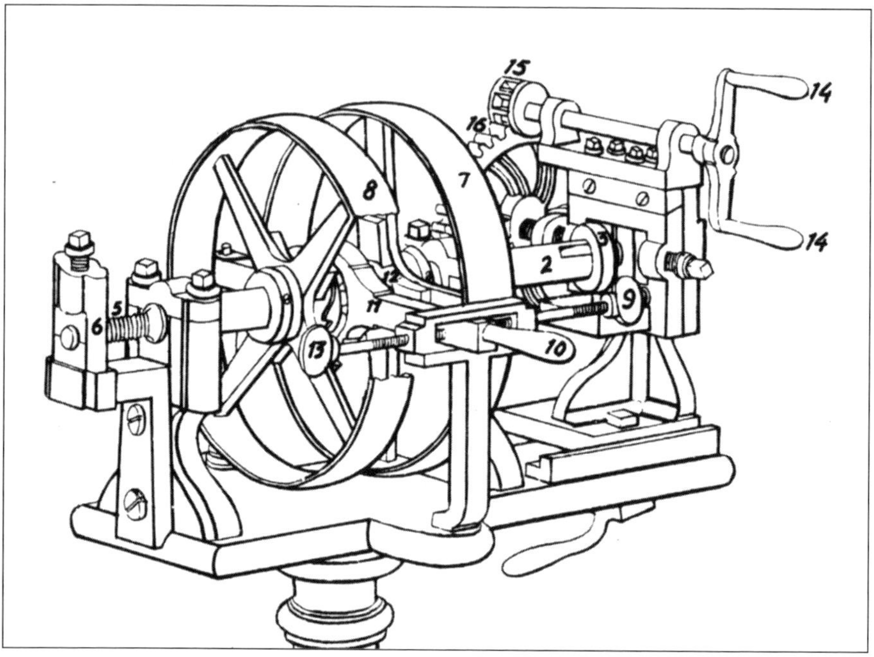

Fig. 4.46. Breech plug threading machine. (Gamel, 1826, Fig. XIII)

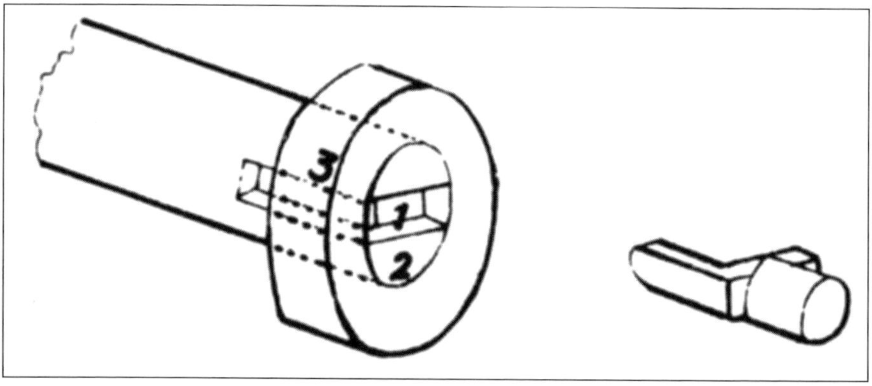

Fig. 4.47. Fixing the breech plug in mandrel nose.

The dies were adjustable by means of the handle, *14*, Fig. 4.46, so the breech plug would be first threaded with the dies partially open, withdrawn by reversing the machine, the dies closed-up to give the finishing cut and the machine again set in motion. The machine was fitted with automatic disengage and reverse mechanisms. The two machines produced one hundred threaded breech plugs per day, as opposed to the twenty-five produced by hand-methods.

Hand methods were still used in the cleaning and adjusting of the threads and in filing the two notches on the breech plug, one on the heel to allow passage of the rear lockplate screw [side-nail] and one on the butt to allow clear connection of the vent with the chamber.

It is noted however that the machines for threading barrel and breech plug could not cut perfect threads and so finishing the thread by hand with taps and dies was resorted to.[33] The barrel then had its outer surface cleaned using the machine in Fig. 4.50, adapted from the honing machine shown earlier and introduced in 1823. Curiously this machine is only described in a footnote and not in the main text.[34]

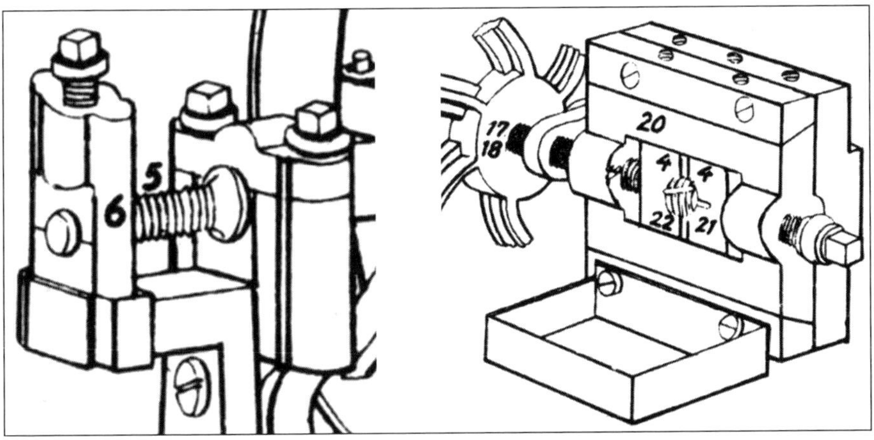

Fig. 4.48. The chaser thread, 5, on the mandrel (left) and the die blocks, right. (Gamel, 1826, Fig. XIII)

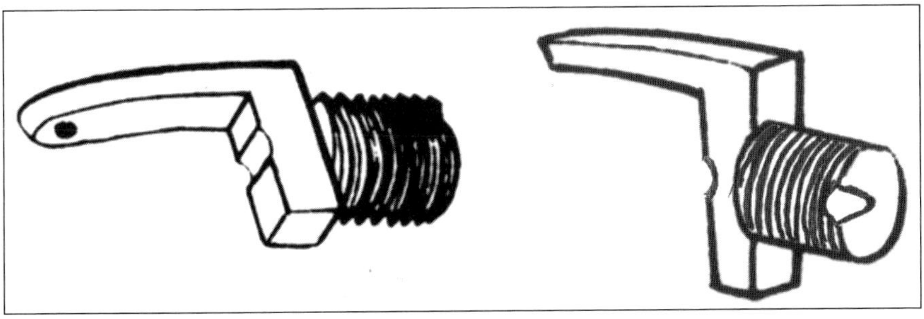

Fig. 4.49. The breech-plug with its two notches filed.

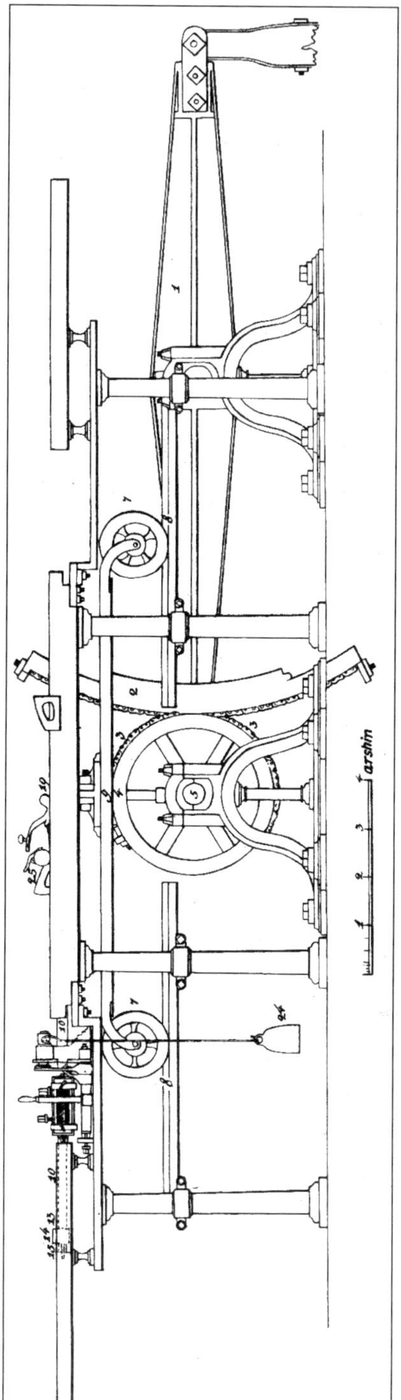

Fig. 4.50. Filing machine for the outside of the barrel. (Gamel, 1826, Fig. VIII)

It operated in much the same way but the ‘honing head’ was replaced with an octagonal ‘filing head’ having concave facets cut with file teeth, Fig. 4.51.

Two of the ‘filing heads’ were fixed to two arms and held against the barrel by weights so that contact was maintained as the barrel was reciprocated between them. It was also rotated by an indexing mechanism, as used in the honing machine, so that the whole of the barrel’s outer surface, excepting the few inches of the muzzle to avoid the bayonet stud and the breech end because of its larger size, was treated. Each machine could accommodate 12 barrels.

Gamel points out that at this juncture the vent was drilled by marking the position on the barrel.[35] Obviously, this would have had to have been done in conjunction with the breech plug *after* it had been screwed into the barrel. Consideration will show that by just laying it against the barrel to gauge the distance of the vent from the breech the chances of the vent being in the right place when the plug was screwed in were very remote, since the threads on plugs and barrels would have had different starting points. Once done, the breech plug would have to remain paired with that barrel. Even when the Enfield Pattern 1853 rifle became interchangeable, the breech plug and barrel were considered as a unit.

The barrel was then ‘proved’ with two shots, the first with six zolotniks (25.5 grams) and the second with five zolotniks (21.25 grams) of powder, each with a single ball weighing seven zolotniks (29.75 grams). Barrels were inspected between each firing and those without defects were transferred to a closed room so that condensed moisture, absorbed by deposits of potassium carbonate, would reveal any cracks. The breech plugs were then unscrewed, the barrels swabbed out with cotton cloth wound on the cleaning rods, and delivered to the inspector where the outside was checked with various gauges and the bore checked with a plug gauge, followed by weighing. Accepted barrels were stamped with day, month, and year of proving. The breech plug was re-fitted and the hole in the tang drilled and countersunk by hand.[36] These procedures are more or less identical to those described by Cotty in relation to the M.1777 musket barrels.

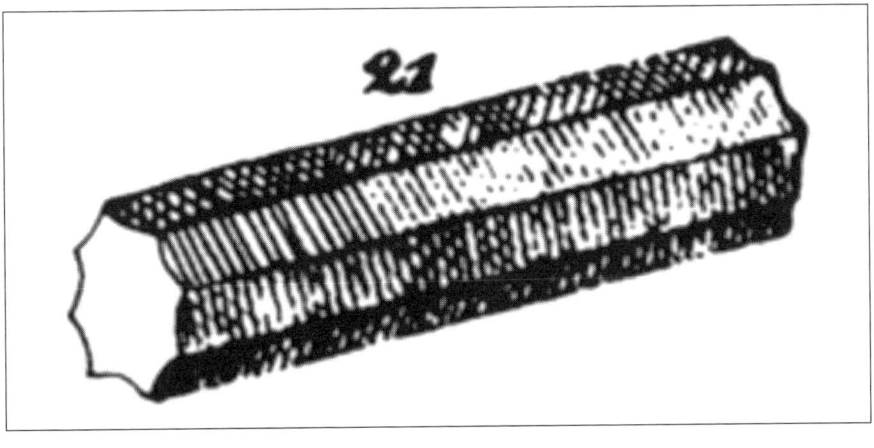

Fig. 4.51. Detail showing filing head. (Gamel, 1826, Fig. VIII)

Work on the barrel was not quite finished. The lug for the bayonet, which had already been brazed on, was then filed to a gauge and the muzzle trimmed to length. It was then filed to fit a gauge of correct diameter and the barrel cleaned with a paste of emery and hempseed oil applied by oak 'cleaning sticks'. This was followed by burnishing with a two-handed, polished, and hardened, steel burnisher which often revealed minor flaws. It was stamped again by an inspector and, with its mating bayonet, delivered into store, ready for stocking.[37]

4.5 External finishing in early 19th century America

Ingenuity was also being applied in America to the outside of the barrel as is evident from Sylvester Nash's patent of 1818[38] for a self-acting lathe for taper turning musket barrels, Fig. 4.52.

The lathe was basically what he described as a *common turning engine* with rack and pinion feed to the carriage, although the patent drawing appears to have

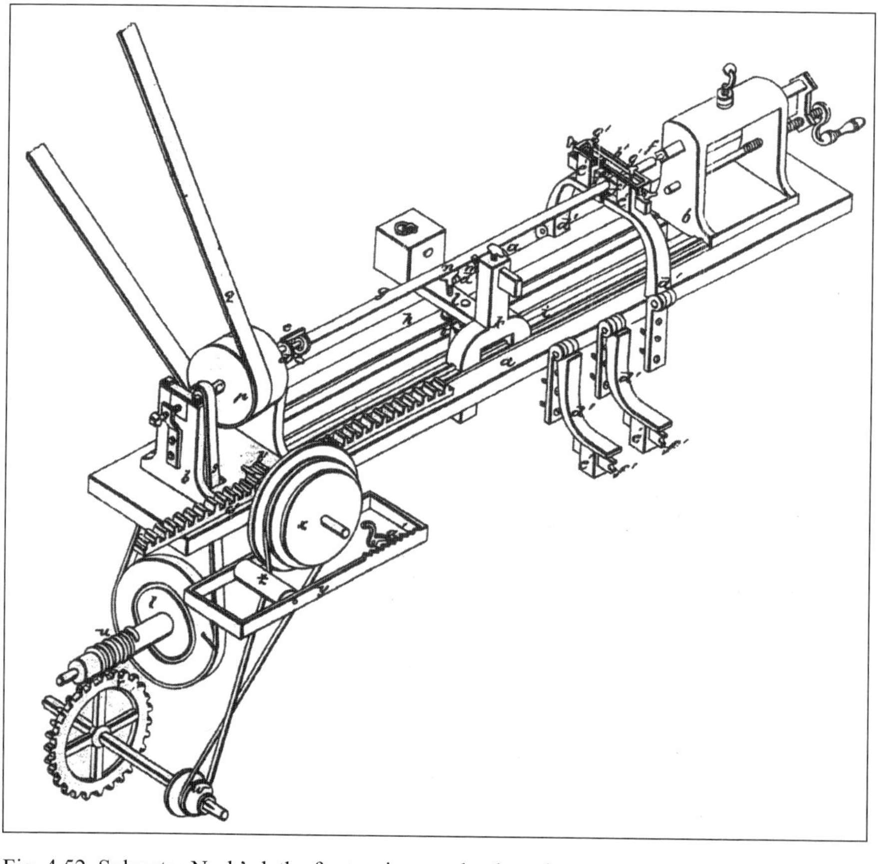

Fig. 4.52. Sylvester Nash's lathe for turning musket barrels.

omitted the linkage between the rack and the carriage, but he had added a number of interesting and novel features.

One of these was the 'bridge' type 'fixed' steadies, Fig. 4.53, which were hinged to the iron bed plate. These supported the 'more slender' parts of the barrel during the turning operation to prevent 'trembling' – what we now refer to as 'chatter'.

The heads of the steadies were equipped with brass 'die blocks' having semi-circular cut-outs which encased the barrel and, with the two halves of the steady in place, the die blocks could be adjusted so they were in contact with the barrel, but not binding it or causing it to deflect, and were held in place by thumbscrews. The heads of the steadies themselves had a rectangular frame fitted over them which carried another thumbscrew, allowing the steadies to be clamped in the correct position. The barrel was fitted with a close-fitting mandrel supported on centres at each end and at one end was attached a 'drive dog' which engaged with the lathe

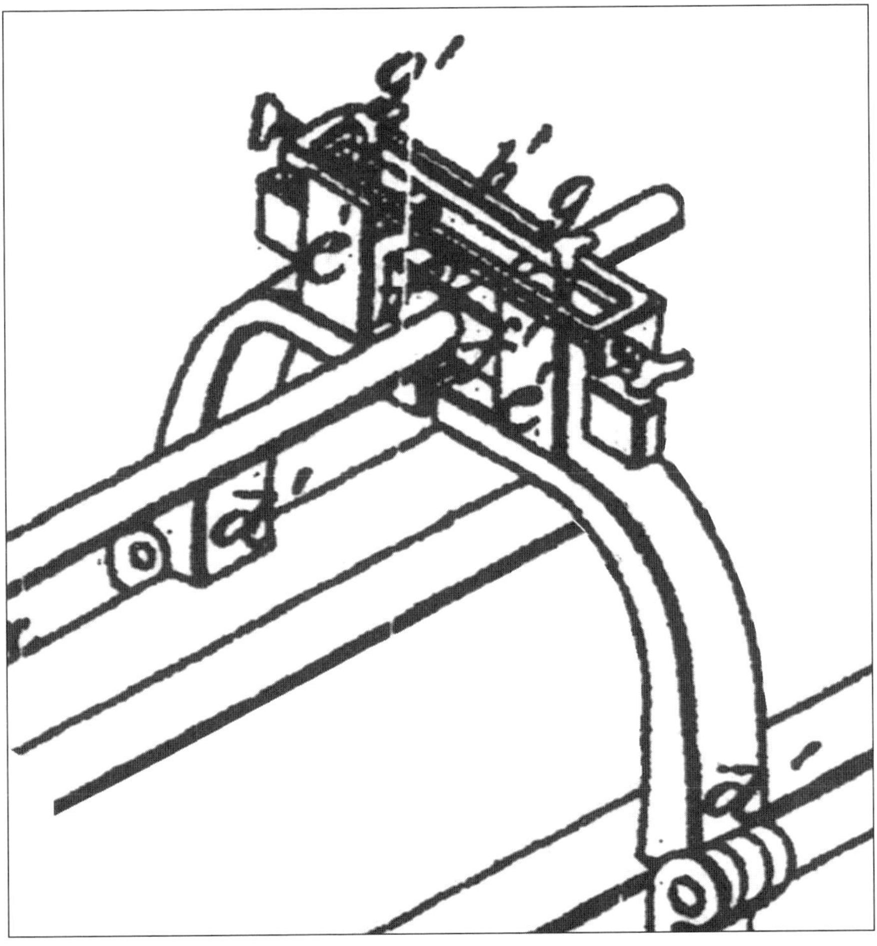

Fig. 4.53. Detail of Nash's steadies.

spindle nose as in Fig. 4.54. Smith[39] contends that this drive dog was fitted to the muzzle but Nash in his patent clearly states *the dog which is fastened to the breech by a screw.*

Examination of the drawing should lead to this conclusion anyway since the steadies would need to be placed where *trembling* would be most likely to occur, i.e., the *most slender* parts of the barrel, which are towards the muzzle.

It has been noted that the carriage was caused to slide along the bed by means of the rack driven by a system of gearing and belts connected to the headstock. If it is assumed that the spindle of the lathe rotates towards the front of the machine where the operator stands, then this drive system presents a curious feature. The worm, or *endless screw* as Nash terms it, appears to have a right-hand thread. It would therefore rotate the wheel in an anti-clockwise direction and if the drive

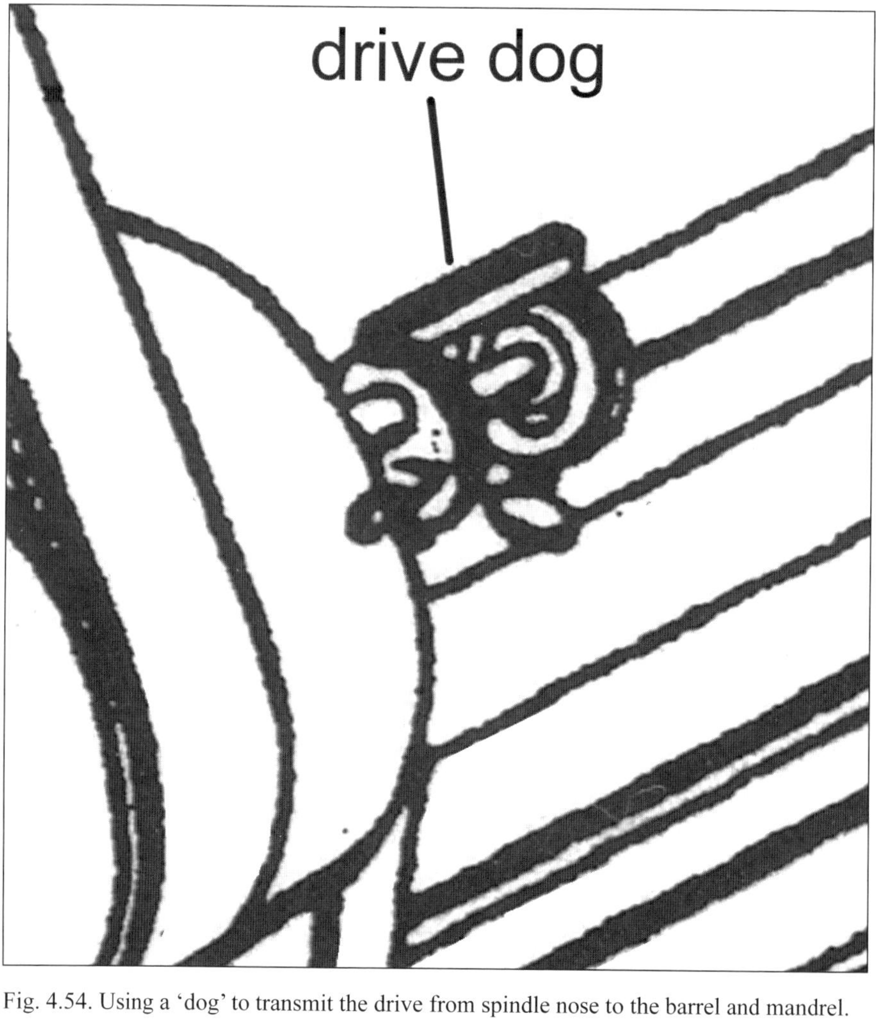

Fig. 4.54. Using a 'dog' to transmit the drive from spindle nose to the barrel and mandrel.

to the pulley wheel, *x*, had the belt in the more usual position, the carriage would move along the bed from left to right, i.e., from the headstock to the tailstock, which would be unusual. However, as can be seen, the belts are 'crossed' to reverse the drive so that the pulley driving the pinion engaged with the rack operates in a clockwise direction and causes the carriage to traverse in the more usual direction of right to left.

The carriage was guided during its traverse by a raised triangular steel fillet attached to the top face of the front of the bed, a feature which became standard practice amongst many precision lathe builders into the 20th century.

To obtain the taper, an iron bar with inclined top face was fitted to the rear of the lathe bed, which bore the bottom face of the so-called 'gauge screw', Fig. 4.55, such that, as the carriage moved from right to left, its rearward projecting *tongue* was progressively lifted whilst contact with the inclined face was maintained by weights applied to the *tongue*.

It should perhaps be pointed out that while in Fig. 4.55 'gauge screw' appears to be in contact with the underside of the barrel, this was not the case in actual practice.

To anyone familiar with lathe work, close scrutiny reveals what appear to be two distinct flaws in Nash's arrangement if the patent wording and drawing are interpreted correctly. The first is that the turning tool, in normal lathe-work, operates on the front face of the workpiece at centre height: the question which arises is how this arrangement can work if the carriage, in moving from right to left to taper the

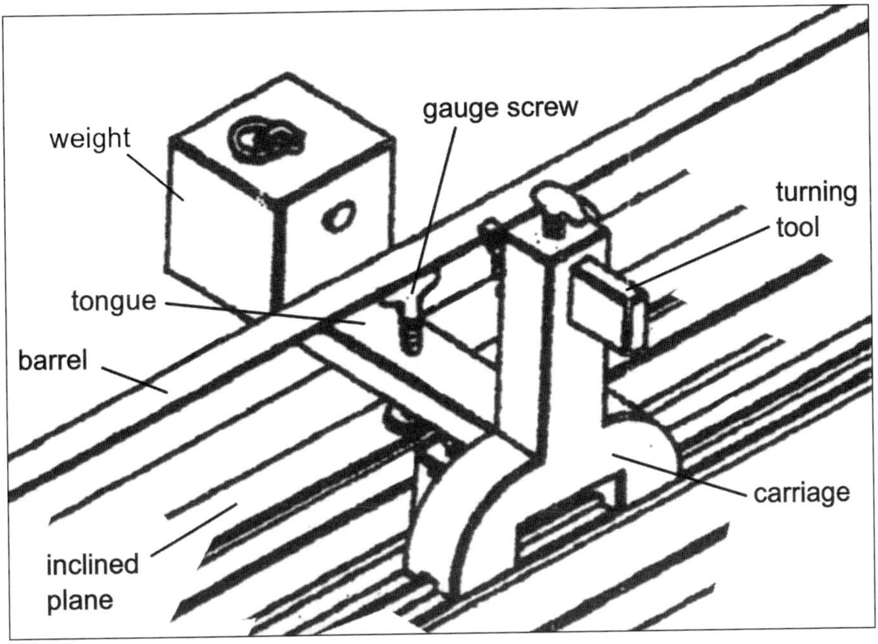

Fig. 4.55. Detail from Nash's patent drawing showing the carriage and its 'gauge screw' with a turning tool bearing upon the barrel.

barrel from muzzle to breech, the gauge screw bearing on the inclined plane, and therefore the tool also, is lowered by the gauge screw bearing on the inclined plane? This will cause the cutting edge of the tool to fall below centre height, Fig. 4.56. If the barrel was turned from breech to muzzle, then the cutting tool would rise above centre height with its attendant problems. For this reason alone, it is suggested that Nash's lathe as shown could not have achieved its designated purpose.

The second consideration applies to taper-turning, more commonly carried out by moving the cutting tool horizontally and allowing both the centre height of the tool to be maintained throughout its passage along the outside of the barrel and for a definite taper to be more easily predetermined. It would seem that Nash's lathe would have worked better with the turning tool mounted *below* the centre line of the barrel.

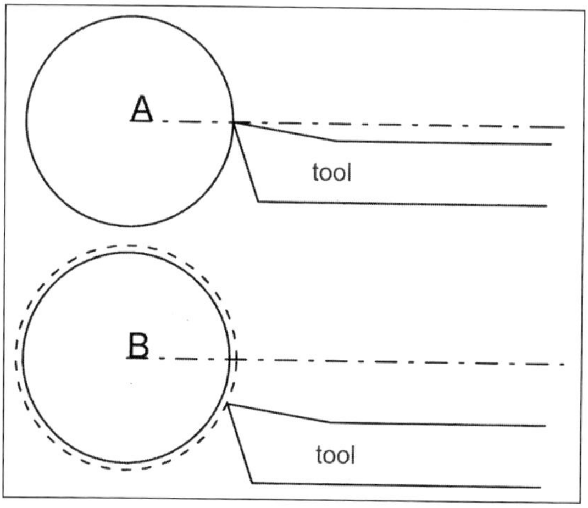

Left: Fig. 4.56. Sketch showing the perceived faults in Nash's lathe. (© P. Smithurst)

Below: Fig. 4.57. Barrel grinding at Springfield Armory. Note that the barrel is rotated on a mandrel and the stone is within a protective guard. (Abbott, 1852, p. 152)

Turning was not the last operation. As late as 1852, at Springfield Armory[40] the barrels were then ground, to remove the tooling marks left by turning, in an almost identical manner to that used in France.

4.6 External finishing in early 19th century Britain

Little is known of the methods used in Britain. In the later 18th century through to the 1840s, various attempts were made at the Enfield Lock barrel mill to replace grinding with lathe turning and a detailed study of this initiative has been made.[41] It only met with limited success, partly because of deep-rooted traditional methods – grinding – in which skilled craftsmen could carry out the work with a sufficient accuracy and speed born of long experience. Such was the importance of grinding that it persisted both at Enfield and even more so in the private trade for many years. Nevertheless, during this period various attempts were made to devise suitable machines and one is covered in a patent issued to James and Jones in 1811.[42] What is of interest, apart from the technology, is that it is believed that this is the same John Jones who was shortly afterwards recruited by Tula. No illustrations are included in the patent, but the description has been used to create simplified illustrations of the principle proposed.

Three substantial (cast?) iron bars were rigidly secured to a framework and arranged about an imaginary centreline of the barrel. Two were in the same horizontal plane and the third positioned above the barrel centre line. The two lower ones were angled so that their inner faces matched the taper required on the barrel and the upper, third bar, had its underside angled likewise, Fig. 4.58.

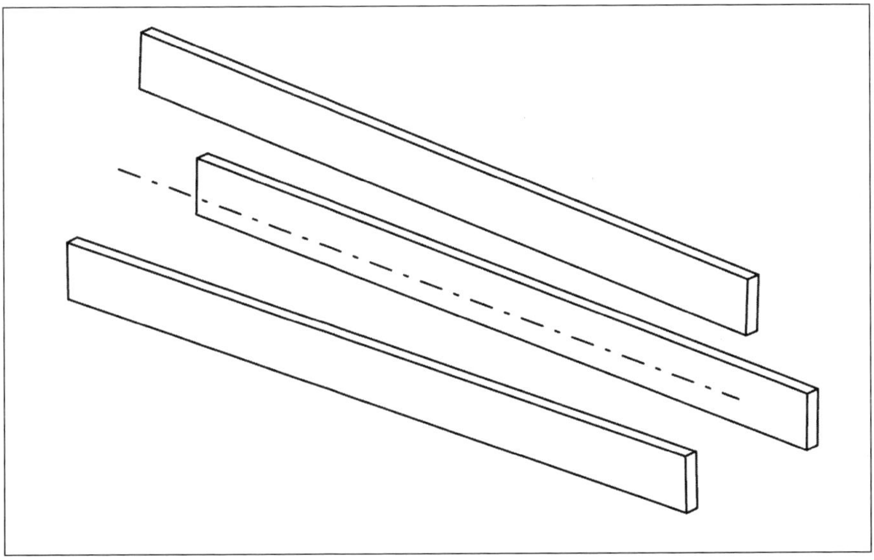

Fig. 4.58. Schematic of James' and Jones' arrangement of the guide bars. (© P. Smithurst)

The two lower bars acted as the lathe bed and at their left-hand end would have been the lathe headstock with a spindle rotated by a suitable power source. At the opposite end would have been a tailstock. Fitted to these bars, and able to slide along them, were three 'carriages', each driven by a feed screw geared to the headstock so that they traversed the guide bars at the same rate. The barrel, having been bored and straightened, was fitted with an accurately made mandrel which was secured upon it and the mandrel with the barrel was in turn, secured at its breech-end to the headstock spindle and at the muzzle end to the tailstock. The general arrangement is shown in Fig. 4.59.

The front carriage was to carry two cutting or turning tools, set close together, the first to perform 'rough turning' and the second taking a finer cut to perform 'finish turning'. The rear and top carriages carried blocks which were set to bear upon the machined part of the barrel immediately behind the cutting tools to act as 'steadies' to prevent the barrel from deflecting upwards or rearwards in response to the forces upon it created by the machining process.

W. Greener writing in 1841, but with particular reference to the sporting gun trade, makes the comment that: *Many barrel makers turn their barrel entirely, by self-acting lathes, and thus obtain a correct taper from breech end to muzzle.* He goes on claim that: *To obtain truth,* [it is believed he is referring to a uniform wall thickness] *it is important that they should be turned.*[43] He notes that the way of fixing them in the lathe is by inserting tightly fitting plugs in muzzle and breech, the plug in the latter projecting and having a carrier fitted to it. These plugs would have

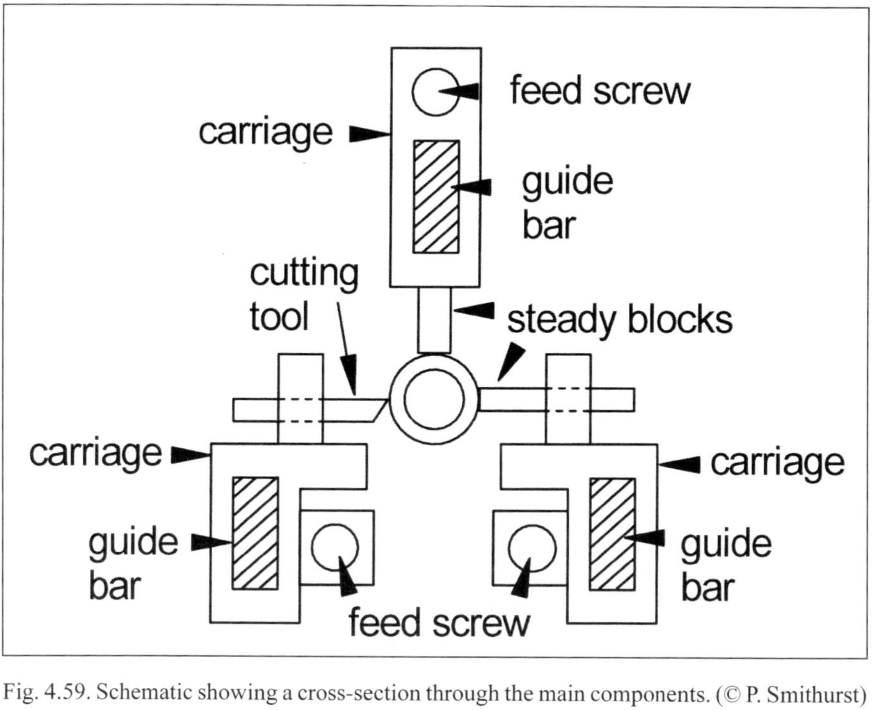

Fig. 4.59. Schematic showing a cross-section through the main components. (© P. Smithurst)

needed conical 'centre holes' in their exposed ends, allowing them to be mounted on matching centre points at the tailstock [muzzle] and headstock [breech], and the carrier on the breech end engaging with a 'drive dog' on the headstock so that in operation, the barrel would be rotated. He then continues by giving a very simplistic and, to anyone having some familiarity with lathes, misleading description of the next stage of the operation: *The leading screw that travels* [i.e., moves] *the slide-rest, is then set in the angle the barrel is to be turned to* . . [author's underlining]. In any lathe, the slide rest, or carriage, has to follow the bed of the lathe. Any attempt to angle the leadscrew would lead to the wrecking of the machine – unless the bed of the lathe, by whatever means, could be independently angled to the centre line of the barrel. It has already been noted that such a geometry, admittedly fixed and not adjustable as in Greener's comment, had already been patented in France in 1787 and more recently by James and Jones in 1811. Is it conceivable that Greener was providing a very inadequate description of one of these machines actually in use? He obviously witnessed, or was told something, but that important question remains unanswered at this time.

4.7 Examining finished barrels

Once a rough barrel tube had been made, their very nature dictated that all the future boring and turning operations remove unequal amounts of metal internally and externally before arriving at a finished barrel. What may have been balanced stresses in the rough tube then became unbalanced stresses in the machined tube, causing the barrel to *warp*.

According to W. W. Greener,[44] before 1795:

> *all gun-barrels were comparatively crooked, there being no reliable method of ascertaining when the barrels were straight . . . but about that year a barrel maker of Birmingham named Parsons introduced a plan of straightening barrels from the inside. His method consisted in stretching a string or fine wire inside the barrel from end to end, and touching the side at each end. He then hammered that side of the barrel until it touched the string.*

However, no priority can be granted to Parsons since this is exactly the same method described as being used in Portugal in the early 18th century, where a *fine viola string* was employed, but it is perhaps understandable that Greener and Parsons were ignorant of this. It was also the method used in France.

Greener then goes on to state that only a few years later the method often referred to as 'shading' was developed. This involved holding the barrel pointed upwards towards a window and causing the shade of the upper edge of the window to move up and down the tube. This was greatly aided by the fine bright finish in the bore. Any irregularity in the outline of the shade indicated where any inaccuracies existed. This, and the straightening of any slight curvature in the barrel, was achieved by laying it on a hollow anvil with a dished face and striking it at precisely the correct point and with exactly the correct force using a wooden mallet - a very highly skilled job.

The identical process was carried out in the United States except that in one recorded instance[45] at Springfield Armory, instead of pointing the barrel upwards at a window, it was pointed downwards at a mirror, Fig. 4.60. The mirror reflected the window which had a diagonal line marked upon it. This took the place of Greener's 'shade' of the upper edge of the window. Having the barrel pointing downwards at a mirror was simply a way of avoiding fatigue of the 'setter' by not having to lift barrels to head-height countless times. Any curvature detected would be corrected in the same fashion as its English counterpart.

Fig. 4.60. Inspecting barrels at Springfield Armory. (Abbott, 1852, p. 150)

Chapter 5

Manufacture of the Lock

5.1 Introduction

There is, moreover, no art in which there is so great a need for tools as that of gunmaking, the perfection of which is always due to the number and perfection of the tools which cannot be better made than by the gunmakers themselves, which requires on their part much more skill than to make a complete lock.[1]

Or, as the renowned English lockmaker, Joseph Brazier, noted:

> *A man who can file gun locks can file almost anything else.*[2]

For the soldier, then as now, the keys to a successful weapon were functional simplicity, reliability, sturdiness, and low cost. At the time of the matchlock and wheel-lock systems, warfare was still very much the province of the *armes blanche* - swords, lances and other edged weapons. To make firearms the primary weapon of warfare, something was required which combined the simple ruggedness of the matchlock with the ability of the wheel-lock to generate sparks. These qualities were provided by the flint-lock in which the ability to create sparks by striking flint against steel was utilised.

The predominant system of ignition throughout the 18th century and into the first quarter of the 19th century was, then, the flintlock. The lock, having a number of components interacting together in a 'train' in conjunction with springs and incorporating safety features, constituted the most complex part of any firearm. It was the only true 'mechanism' in a firearm in the periods being studied and constitutes, along with the trigger, the means by which the gun is fired. For those not familiar with flintlock muskets, the 'lock' components are shown in Figs. 5.1 and 5.2 along with their English terminology.

It is thought that this form of lock having a sear engaging with a tumbler with *two* notches or *bents*, but without a bridle to support both tumbler and sear, is of French origin.[3] The 'double bridle', which supports sear and tumbler, seems to have appeared around 1740[4] but from where it came is not recorded.

Such muskets were mostly loaded from the muzzle with a pre-defined charge of gunpowder and a musket ball contained in a paper 'cartridge'. One end was torn

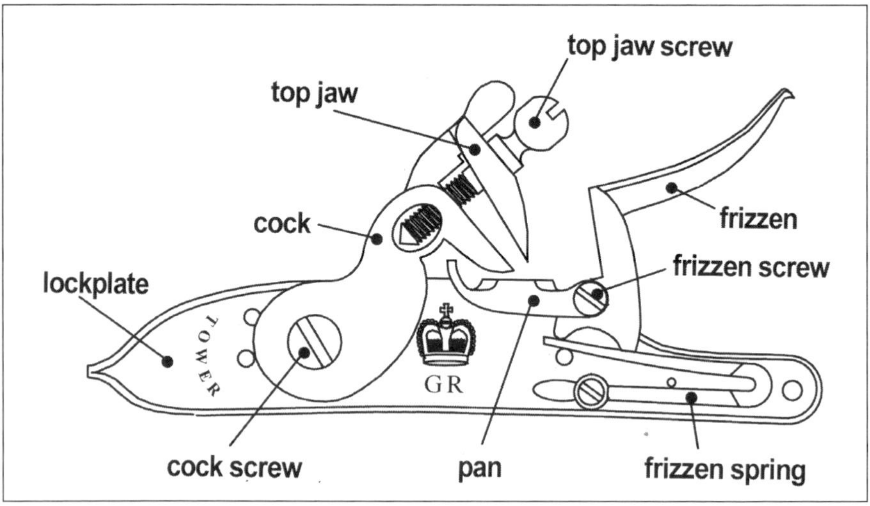

Fig. 5.1. The outer features of the flintlock (© P. Smithurst)

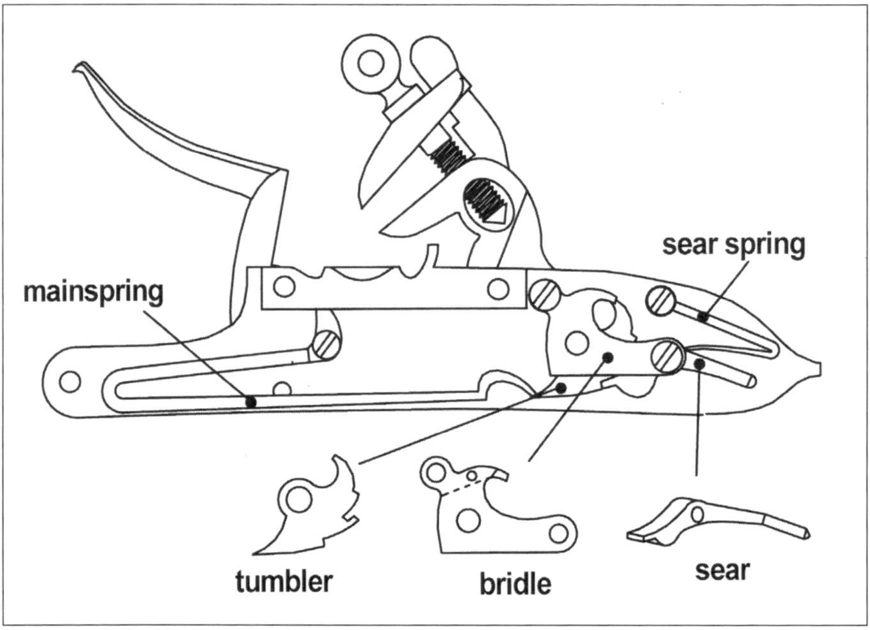

Fig. 5.2. The inner features of the flintlock (© P. Smithurst)

off, a small amount of the powder was poured into the hollow 'pan' and the frizzen was closed on top of it and held in place by the frizzen spring. The remaining powder was poured down the barrel, followed by the musket ball. The ball was a loose fit in the barrel and was usually prevented from falling out by the remains of the paper cartridge being rammed down on top of the ball which also seated it upon the powder – an air gap between powder and ball could result in a burst barrel when fired!

The lock components acting in a 'train' are shown in Fig. 5.3.

It will be seen that the 'hook' tip of the long branch of the mainspring rested on the forward tip of the tumbler. Both of these contact surfaces were polished and hardened to reduce friction and wear. The tumbler was held in place by the bridle which provided a support for the inner pivot of the tumbler. The outer, larger, pivot of the tumbler passed through the lockplate and terminated in a squared portion onto which was fitted the cock by means of its square hole. The tumbler was also in constant contact with the forward tip of the sear, also pivoted, through the action of the sear spring. A projection at the rear end of the sear, the 'tang', was acted upon by the trigger mounted in the stock, and pivoted on a fine pin. All the components of the lock had to be both accurately made to ensure they interacted correctly and made from material that withstood the stresses they were subjected to, so that the musket would function correctly.

In operation, a piece of specially shaped flint was clamped in the jaws of the cock. When the cock was pulled back from its 'resting', or 'fired' position, shown at *1* in Fig. 5.4, against the pressure of the mainspring and partially rotating the tumbler, at a set point the sear, also acted upon by a spring, fell into a shallow undercut notch on the tumbler, shown as position *2*.

The sear in position *2* had its nose tucked into the undercut and resisted any attempts to move it by pulling the trigger – for this reason it was known as the 'safe' position and was used when loading.

Pulling the cock back further disengaged the nose of the sear from the undercut and eventually it dropped into a plain step on the edge of the tumbler. This was known as the 'full-cocked' or 'firing' position since pressure on the trigger would disengage the sear, allowing the tumbler to be rapidly rotated under the action of the mainspring, carrying with it the cock fitted to it. This brought the piece of flint

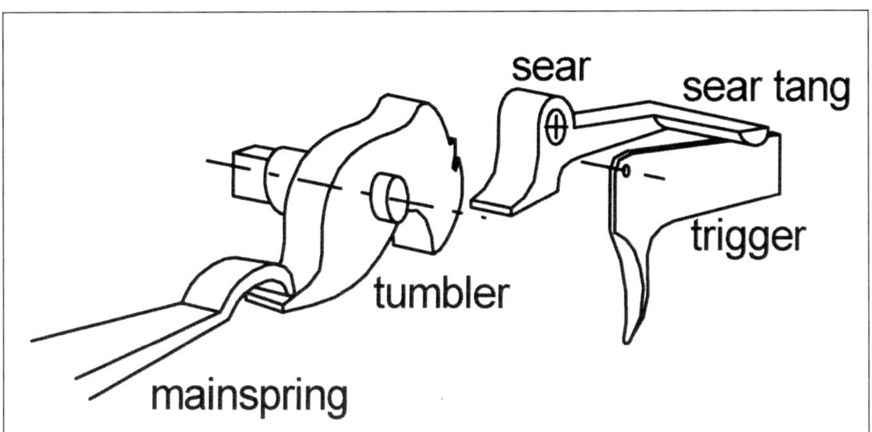

Fig. 5.3. The 'train' of interacting components. (© P. Smithurst)

into forceful contact with the hardened steel face of the frizzen, hinging the frizzen out of place to expose the 'priming' powder in the pan. At the same time, the piece of flint sliding down the hardened face of the frizzen created a shower of sparks which were directed into the pan, igniting the priming powder.

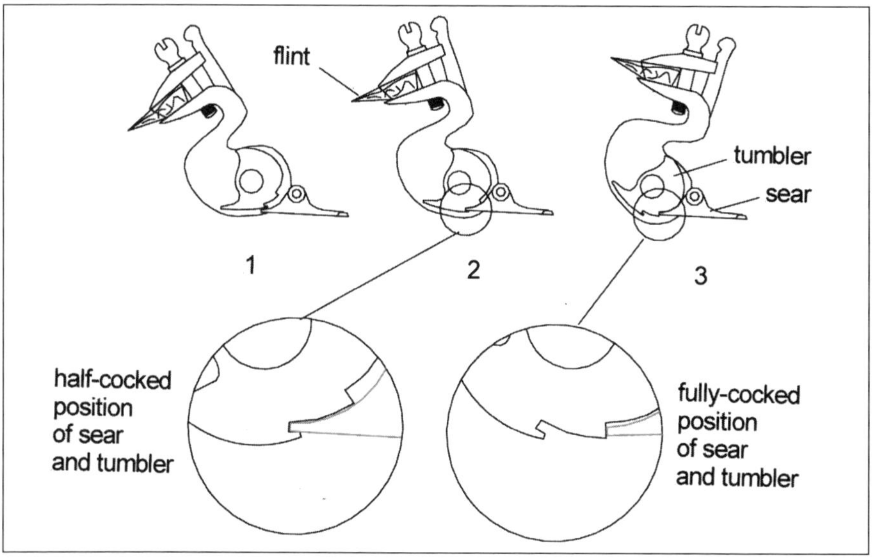

Fig. 5.4. Motions of the flintlock in action. (© P. Smithurst)

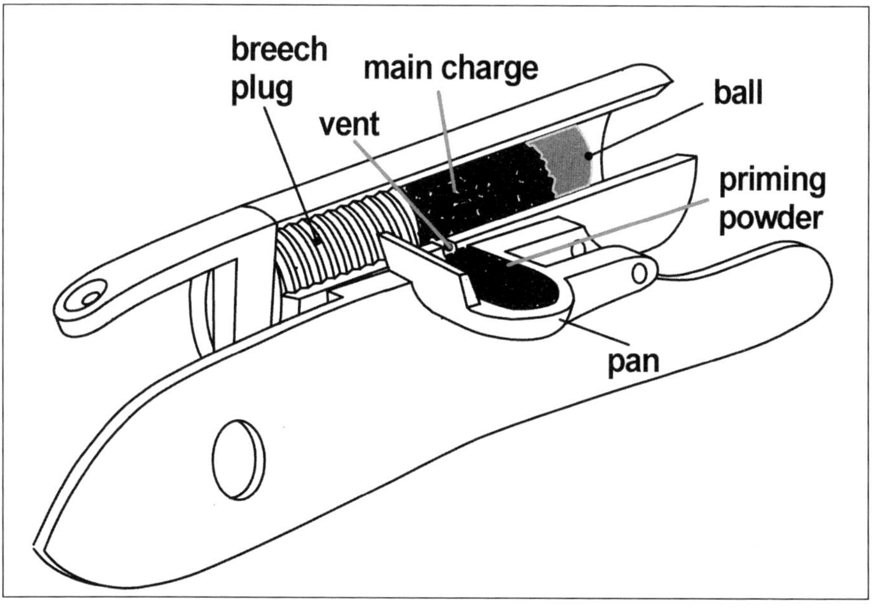

Fig. 5.5. The configuration of lock and barrel. (© P. Smithurst)

When the musket was assembled, the lock lay alongside the barrel so that the pan of the lockplate, which contained the priming powder, lay alongside the 'vent', commonly called the 'touch-hole' in the barrel, which communicated with the main charge in the barrel.

Thus, the flame from the burning priming powder passed through the 'vent' and ignited the main charge in the barrel, 'firing' the gun.

5.2 Lock making in France c.1777-1806

Unfortunately, the earliest detailed account of gunmaking, *Espingarda Perfeyta,* while commenting on the requirements of a good lock, offers no detail on how locks were made. It does, however, indicate that a gauge was used to check the overall shape and dimensions of a lockplate.[5] Gordon[6] notes that the French savant René Antoine Ferchault de Réaumur, in his 1722 treatise on iron and steel,[7] explained the benefits of having interchangeable parts in firearms. Interestingly, he also hinted at the process later developed by Henry Bessemer in his pursuit of a method for converting molten cast iron into molten wrought iron for artillery barrels, resulting in what became known as 'mild steel'.

In 1723, a year after Reaumur's suggestion, Guillaume Deschamps, a French armourer *cum* inventor, produced 660 interchangeable locks as judged by inspectors appointed by Jean-Florent de Vallière, Inspector General of Artillery. While de Vallière recognised the value of a standardised lock, he rejected Deschamps' interchangeability, achieved through extensive hand filing, as *superfluous, impractical and expensive*.[8]

It was not until the later 18th century that details of lock making began to emerge and relate to the common form found on muskets and thus needed to be manufactured economically and in quantity. Moreover, it was a period which witnessed momentous changes in France. In 1776, General Jean-Baptiste Vaquette de Gribeauval became Inspector General of Artillery and implemented a scheme of standardisation to ease both manufacture and the supply of all the necessary equipment of an army in the field. He is most noted for his new design and standardisation of field guns and carriages, but small-arms also came under his control, and it was one of his workers in this field, Honoré Blanc, who was to take standardisation to new levels through the pursuit and achievement of a concept which is vital to many manufacturing industries today – interchangeability.

Gribeauval's death in 1789 was closely followed by the tumultuous French Revolution during which many of the aspirations associated with the previous regime were discarded. Thus, in the closing years of the 18th century, two competing manufacturing technologies emerged: the 'high-accuracy' and partly mechanised version of Blanc was replaced by hand-craft techniques favoured by the revolutionary governments and which, because Blanc's approach was very much a novel initiative, are likely to reflect practices elsewhere. Blanc's work is summarised in a memorandum he submitted to the National Assembly in 1790 and was reported upon by a committee composed of members of the Royal Academy of Sciences in 1791. Translated transcripts of these can be found in Appendix 2 and passages quoted from them are not referenced.

From his own testimony Blanc first entered the sphere of arms manufacture in 1748 and, having spent some time as Master Armourer at the Charleville factory, in 1763 he was appointed *côntroleur de la manufacture d'armes* at St. Etienne. There he designed and created all the firearms issued in 1764, 1766, 1768 and 1774,

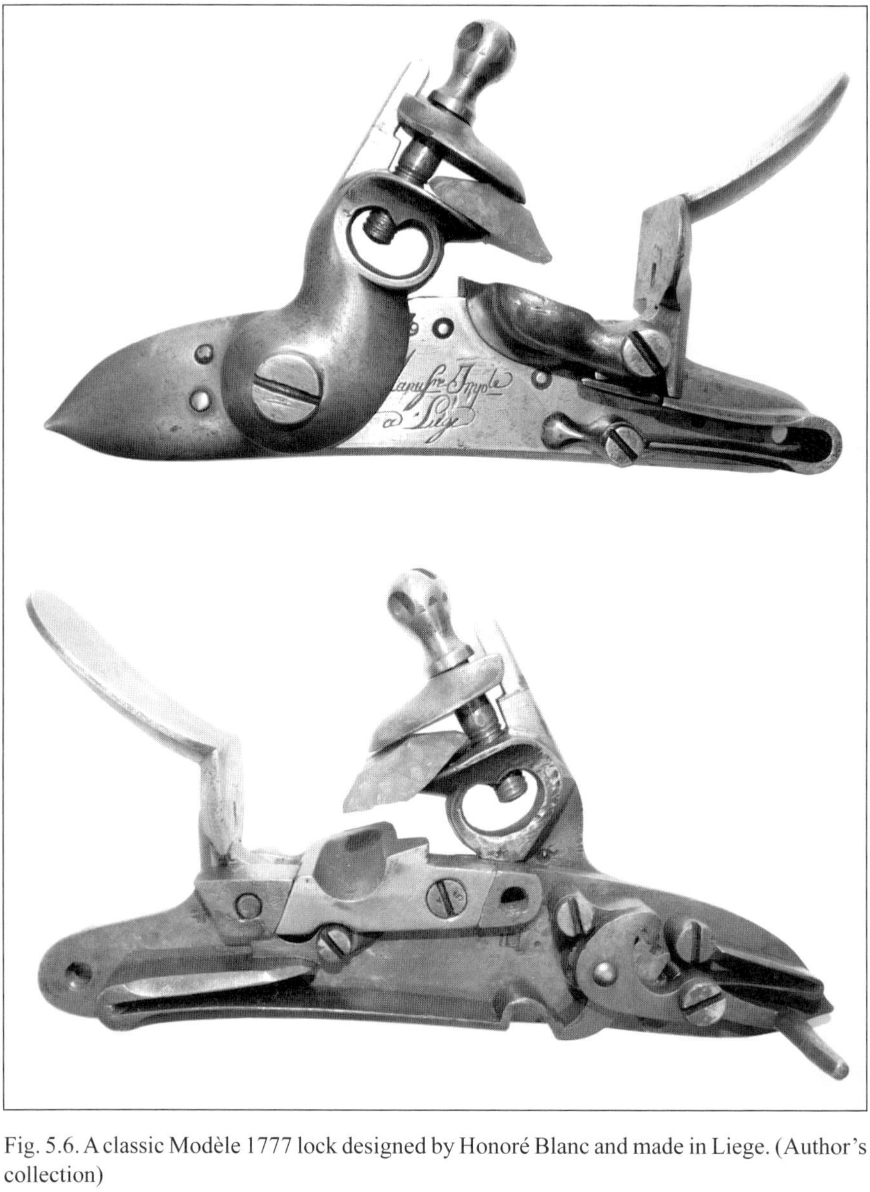

Fig. 5.6. A classic Modèle 1777 lock designed by Honoré Blanc and made in Liege. (Author's collection)

culminating in his iconic *Modèle 1777 Fusil d'Infanterie* musket and bayonet which became the standard pattern on which French military firearms were based.

Gribeauval probably became aware of Blanc's work through his 1777 Musket and, in the following year, 1778, Blanc was appointed Principal Controller of the other three arms manufactories in France, in addition to St. Etienne. After visiting the factories at Charleville, Maubeuge and Tulle in 1782, his attention was drawn to *the multiplicity of imperfections which negligence had left in the manufacture of arms* and he *gave himself up to every reflection capable of suggesting to him the means of remedying it.*

As a result, he created a plan of work which embraced all the parts of the weapon and *occupied himself with the invention of various measures, tools, machines and dies, hitherto unknown in our manufactures, in considerable number to facilitate the work of more than 800 workers, and to give to this work all the precision possible* and introduced standardisation by specifying the proportions and dimensions of the weapons and their components.

These alone constitute a major change in the approach to manufacturing a composite and complex product. It assumes even greater importance considering that some workers, such as the lock makers, often worked from their homes at some distance from the manufactory itself and the only means of ensuring their work conformed was by the use of matching dies, jigs and gauges. Another image of the set of gauges for the M. 1777 musket made by Blanc in the period 1777-1795, is shown in Fig. 5.7.

Around 1786, Blanc's work came to the notice of someone he refers to as *M. Châtelet*, but who was likely to have been the *Duc du Châtelet*, a prominent member of the Comité Militaire. Like Gribeauval, he was impressed with Blanc's work and aspirations and intervened with the Minister to give Blanc space at the fortress at Vincennes to develop his ideas. There, he devoted his efforts to the manufacture of locks which resulted in two documents of primary importance to the study of the early pursuit of Interchangeability. The first is Blanc's *Mèmoire Importante* which he presented to the National Assembly in 1790, reporting particularly on the manufacture of gun locks with interchangeable parts and their utility:

> *that the pieces composing a lock being manufactured separately, may, after they have been completed, be intermingled and confused together, and then each taken at random to serve to form the lock, by means of which there is no-one who is not able to reassemble a lock, with the same ease as if it were only a question of changing the flint.*

and

> *. . . one would enjoy the inestimable advantage of carrying in a crate, spare parts of all kinds, hardened and tempered, polished, and ready to replace a broken or missing piece, since it would no longer be necessary to have an armourer in charge provided that they* [the soldier] *can dismantle them, which can easily be taught.*

Fig. 5.7. Another image of Blanc's set of gauges for the M. 1777 musket. (Inv. No. 2013.0.11. Photo (C) Paris - Musée de l'Armée, Dist. RMN-Grand Palais - © Emilie Cambier)

The second is a report in 1791 by Messrs Borda, le Roy, de la Place and Coulomb of the *Academie Royale des Sciences* on Blanc's claims and which endorsed them in no uncertain terms. Blanc sent another petition to the National Assembly which resulted in the War Office appointing a commission of artillery officers to investigate his claims and which echoed the Academy's findings.[9] This was on the eve of great upheaval in which the ebb and flow of political ideals in the French Revolution had an impact on the way arms were produced, and these interactions are covered in excellent detail by Alder.

Blanc, realising that he would not get support from the state, asked to be released from his official post and to be allowed to establish himself as a private entrepreneur with an interest-free loan of 30,000 livres. The commissioners concurred, the government reserving the right to make copies of Blanc's machines, and the Minister approved the project on 3rd April 1792. In November 1793, while the Committee was hastily assembling the Paris Manufactory, he was authorised to purchase the Ursuline convent in Roanne, along with a watermill. In return, Blanc was to manufacture thirty-thousand interchangeable locks per year.[10]

How did Blanc make his interchangeable locks? His mèmoire is less informative on the finer details and is more concerned with presenting an overall picture of the then state of arms production in France and what his proposals would accomplish. The *Academie* was more intent on establishing whether Blanc's claims in relation to the production of locks with interchangeable components were accurate and to establish whether the system he was proposing was viable. Their report contains a little more detail of some manufacturing processes but still falls short of describing them all. Much of this shortfall in detail is covered by Cotty, despite his dismissal of Blanc's claims. Through a combination of these documents, and drawing upon other sources, it is possible to arrive at a general understanding of the manufacturing processes.

Blanc claimed that he *invented for this purpose several machines and a large number of tools of such usefulness,* but unfortunately, he does not elaborate.

He later states his inventions had *the essential object of making the parts of locks, and to which we give the definitive form with machines or dies, the regularity of which cannot vary, and this cannot be expected from the hand alone of the workmen.*

This expresses the essence of the use of machines which ultimately create greater accuracy, and assuming the accuracy of such machines, lead almost naturally to interchangeability. His mention of 'dies' is some indication of at least one preliminary method of making the components. Indeed, Cotty gives Blanc the credit for being the first to apply die stamping to creating lock components[11] although, as will be noted later, this may not be entirely accurate.

The Lockplate

The Report of the Academicians is a little more forthcoming. Despite the comment:

> *In order not to load this account with superfluous details, we shall first speak of what the gunmakers call a lockplate and on which are mounted or screwed all the parts which, by their union, compose or form what is called a lock, they proceed:*
>
> *Now this lockplate may be regarded as a sort of plan, on which are holes in invariable relative positions and which define the places where the parts must be fixed. Consequently, it deserves special attention, so that the means by which the author succeeds in making these different holes will be well known, so that in all the lockplates, pierced with the same means, the components all occupy the same places, so their respective positions are identical, which is of primary importance.*
>
> *The lockplate being forged with its two trapezoidal eminences on the sides of the opening for the pan . . . and which, by giving strength to this plate, form, with the pan, what they call the lockplate, is given its specific shape.*

It is useful at this point to illustrate a lockplate of this type to show the 'trapezoidal eminences' referred to in the Report which form abutments for the pan to seat against and to be drilled and tapped to receive the frizzen screw and a lockplate securing screw.

It is possible to assume that dies would be used in this process since, as noted above, Blanc refers to them in his Memoire, and they would certainly have provided a faster method than the traditional method of forging lockplates by hand, followed by extensive filing to bring them to size and shape. However, the abandonment of such mechanised techniques is perhaps reflected by Cotty when he states[12],

> *The worker forges the pieces by ordinary means; but does not finish them with the hammer with the same care, except however the lockplate and the three springs, which are not stamped . . . the lockplate, because it is iron, must be perfectly clean and soft to be drilled* [author's underlining]
>
> but concedes that, *the forger, to abbreviate the work without affecting the precision, quite often uses in the manufacture of various parts, half-dies, by means of which they give approximate shape to different pieces when the iron has been sufficiently forged.*

Once the lockplate had been forged, it then had to be filed, and as noted in the Report: *For this purpose, it is mounted and filed on another, which is hardened, and which has precisely the form which the former must have when finished.*

In contrast, Cotty states that *one gives the outline to the lockplate by filing it between two filing guides.*[13] Unfortunately, no explanation of exactly how the lockplate was located and fixed securely to the filing jig or guides has been

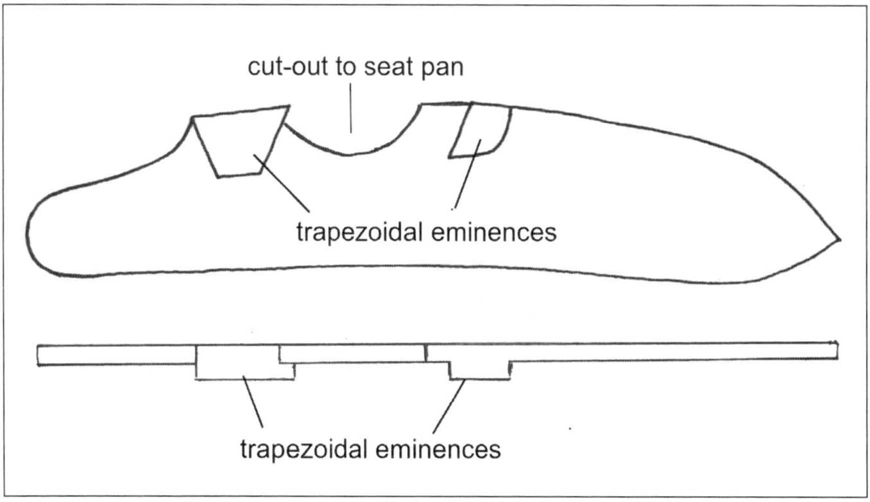

Fig. 5.8. Sketch showing the *trapezoidal eminences* on the inner face of the lockplate. (© P. Smithurst)

found. It is reasonable to suppose that unless it was securely fixed, when the lock and jig assembly needed to be held in a vice in a different position to work on a different portion, there was a risk of disturbing its location. By having a hole, such as the large one for the tumbler, formed at this early stage would have allowed the lockplate and jig to be located and secured by a bolt, for instance. Indeed, this view is supported by a description of operations at the Versailles factory, where the lockplate: *On removal from the stamp it is placed in a machine, the holes of which serve as guides to the different punches with which all the holes are pierced, except those of the rear and front which are pierced by the drill. The first of these holes, that of the arbour of the tumbler is the* <u>*only one which is formed in the stamping*</u>*; the others are punched with a hammer.*[14] [author's underlining]

Here Cotty contradicts his earlier statement that lockplates were not stamped! However, the lockplate would need to be flat on its inner face to sit correctly on the jig and unless the 'trapezoidal eminences' had recesses into which they fitted and were correctly formed beforehand, how was the correct alignment of the lockplate assured? Was the lockplate forging sufficiently flat and these 'eminences' formed with sufficient accuracy in the die-forging operation to satisfy that need? According to Cotty, when it came to inspection, *the inner face of the lockplate must be perfectly flat, as well as the outer surface, except the tail which is rounded.*[15] [In this context, 'rounded' refers to the convex outer face] In the absence of other evidence, it can only be assumed that 'flatness' was achieved by filing.

In 1794 a number of sketches were made to accompany a planned book on gunmaking in the Paris Manufactory and being prepared by Hassenfratz, but which was never published. It might be noted that at the time of his Memoire, Blanc had suggested the creation of an '*Encyclopédie-pratique*' – not a book but a workshop where superior craftsmen could perfect their skills in building machine tools.[16]

One of the Hassenfratz illustrations, Fig. 5.9A, shows various operations being carried out by artisans on the lockplate. One, Fig. 5.9B, shows a lockplate held in a vice. Alder has interpreted the faint caption on the drawing as the artisan filing it to a gauge[17] but no gauge is in evidence.

The Paris Manufactory, created under one of the incarnations of Revolutionary Government, did not have a high reputation for quality. By 1798, control of arms production had reverted to artillery officers and according to Inspector Agoult, the 'Revolutionary No 1 Gun' which the Manufactory of Paris had created in An. II (1795/6), was a *bastard arm . . . it is difficult to determine the nature, dimensions, and price of an arm whose production is arbitrary, and which is composed of so many different models.*[18]

General Gassendi, a friend of Napoleon, went even further by denouncing both the weapons and the administration of the manufactory: *And that is where the marvellous Ateliers of Paris have led us: the weakness of the governors, the ignorance of the supervisors, the rapacious greed of the contractors, the destitution of those armouries where they do know how to make arms . . .*[19]

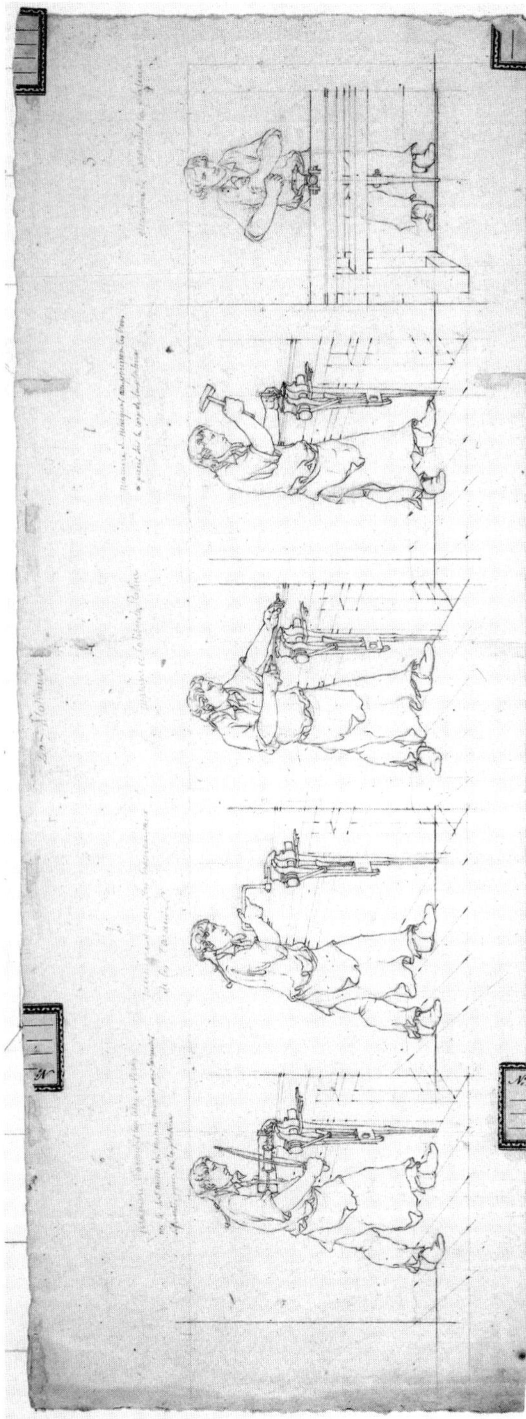

Fig. 5.9A. Lock makers at work, drawing c.1794. (Inv 13571.424, © Musée des arts et métiers-Cnam, Paris/photo Dephti-Ouest)

Right: Fig. 5.9B. Detail from Fig. 5.9 showing a lockplate being filed. (lnv 13571.424, © Musée des arts et métiers-Cnam, Paris/ photo Dephti-Ouest)

Below: Fig. 5.9C. Further detail from Fig. 5.9 showing a lockplate being filed but there is no obvious indication that a filing jig is being used. (lnv 13571.424, © Musée des arts et métiers-Cnam, Paris/photo Dephti-Ouest)

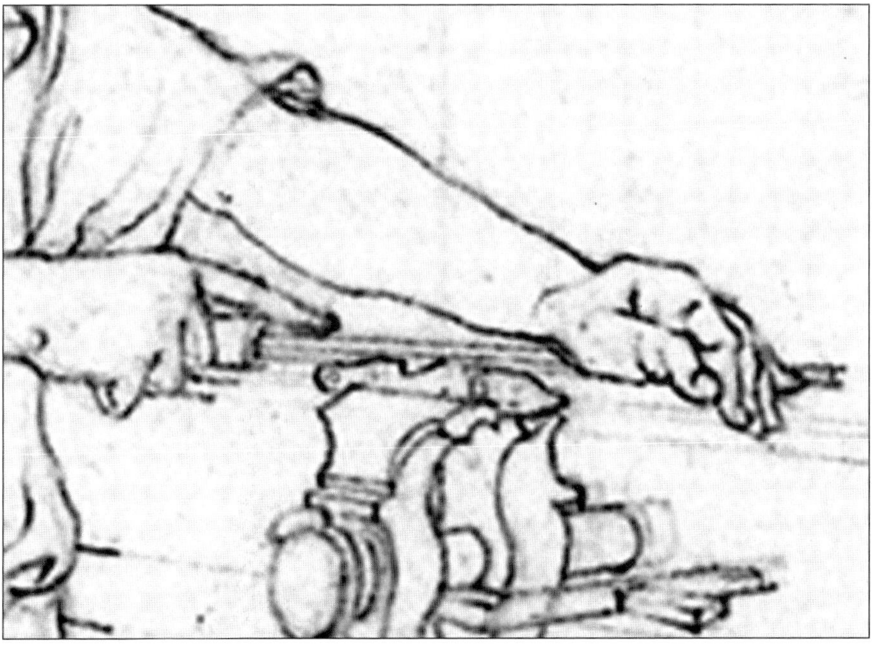

It seems that Blanc's ideals had been overlooked or ignored during this unstable period but, taking a closer look at one of Blanc's gauges, Fig. 5.10, apart from their beautiful finish and inscribed with their purpose, it can be compared with a detail from a copy of a drawing of lock components, Fig. 5.11.

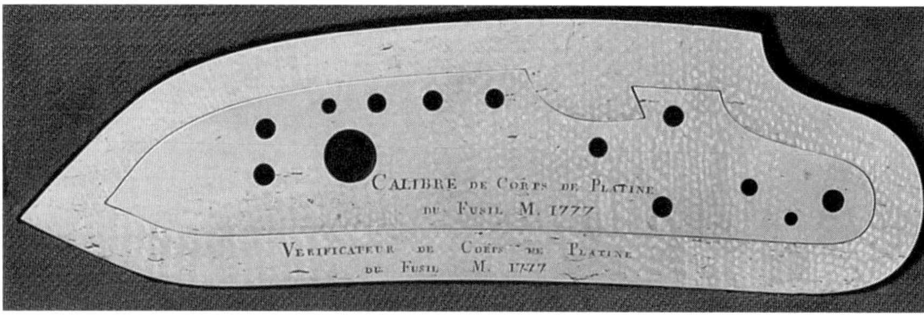

Fig. 5.10. Detail from Blanc's set of gauges showing a lockplate and its profile gauge. (Inv. No. 2013.0.11, Photo (C) Paris - Musée de l'Armée, Dist. RMN-Grand Palais - © Emilie Cambier

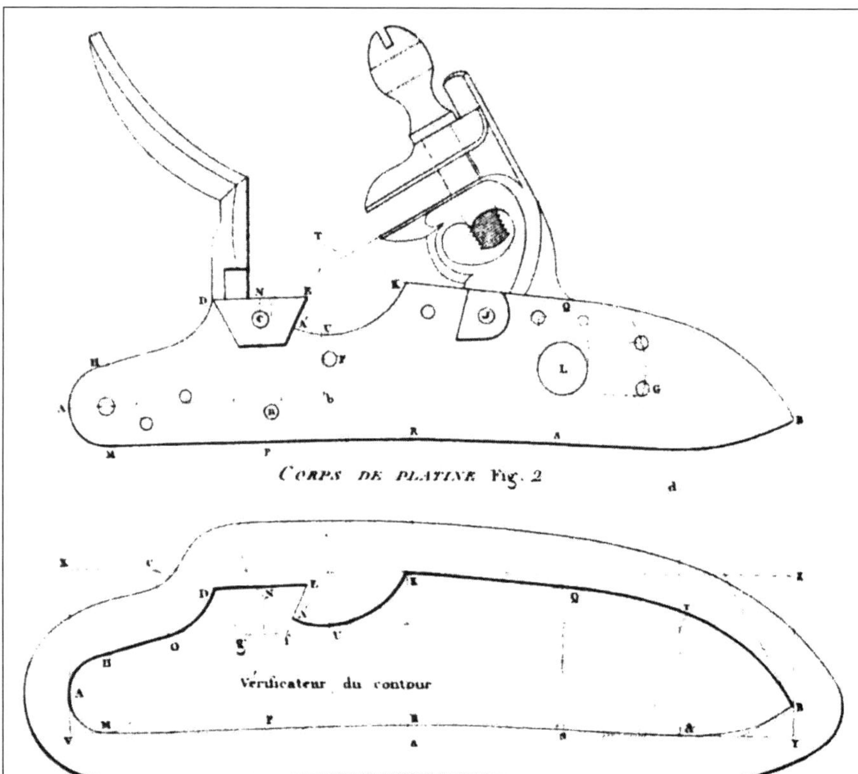

Fig. 5.11. Detail from Official drawing of the Model 1777 An IX (1802) of lockplate and profile gauge as seen from the inside face. (Alder, personal communication)

The original drawing, once in the collection of Service Historique de l'Armée de Terre (S.H.A.T.), was unfortunately lost or stolen according to Alder[20] but he kindly provided me with an old copy. This accurate drawing, showing construction lines, centres of curves, dimensions etc., and which uses standard orthogonal projections, is the earliest of its type encountered depicting firearms components. Its close correspondence with the gauge shown in Fig. 5.10 and its date, taken together with the set of M. 1777 gauges, reflects a resurgence of the French drive for, and approach to, repeatable accuracy in manufacturing which is the foundation of interchangeability.

The accuracy of the filing of the trapezoidal eminences on the inner face of the lockplate was very important since these also formed abutments against which the cast brass pan was located and, being undercut at the front, meant that an accurately formed pan only needed to be secured at the rear by a single screw. The pan itself, Fig. 5.12, has a complex geometry needing to be accurately produced and whilst its initial form would be provided in the casting, its final form, and especially its mating surfaces with the lockplate where a close fit was required, would need to be accurately controlled.

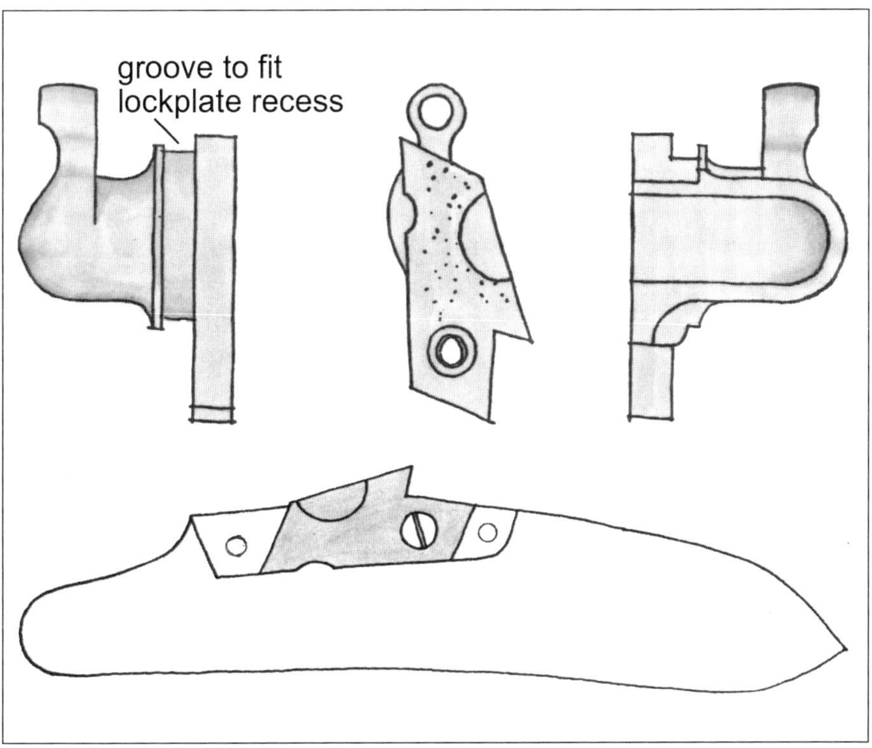

Fig. 5.12. *Top:* Drawing of the pan of the model 1777 musket; *Bottom:* How the pan seated on the lockplate - its close fit with the *eminences* enabled it to be secured with a single screw. (© P. Smithurst)

The official drawing of the pan and its gauge, Fig. 5.13, reinforce the French drive for precision.

How precision was achieved we are not told, but some enlightenment is offered: *It is milled, and the holes are pierced by ordinary means.* [hand drilling?] *All the holes of the other pieces are pierced by mechanical means* [punching, drilling machine?], *and it would be possible after having adjusted the pan to a gauge of the lockplate, to pierce the two holes in the same manner with the help of a guide, which would be easy to apply to both holes, and no doubt there is a method in use at Roanne.*[21]

Cotty's reference to being 'milled', in the absence of any other evidence, relates to the 'hollow' of the pan, and Diderot shows a simple 'milling' machine, Fig. 5.14, using a rotary file-type cutter to perform this task – not a crucially accurate requirement but one still needing care in ensuring the pan's bottom face is centred on the vent.

This same 'machine' is shown in a book on the history of files and filing.[22] The precision with which the pan was formed is indicated by the nature of the gauge and also the brief inspection process (Cotty, 1806, p. 77): *The pan must seat well on the lockplate; its milling must be very straight, and the inclined plane of its lateral edges well squared with that of the lockplate; its total length* [in the direction of the milled hollow] *must be, on all the military lockplates, equal to the greatest width of the lockplate.*

One question arises regarding the pan milling machine – how was the pan presented to the cutter? To have simply been hand-held would have been difficult, not to say dangerous, for the operative and hardly likely to result in accuracy. A mechanically attached and moveable fixture would surely have been required but no evidence has been found to suggest this was the case.

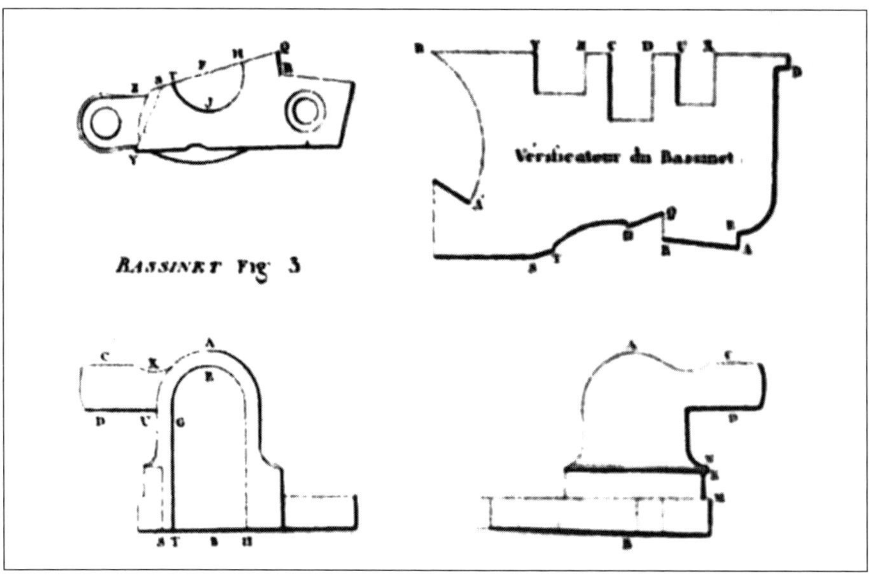

Fig. 5.13. Detail from Official drawing of the Model 1777 An IX (1802) of the pan and its gauge. (Alder, personal communication)

In the absence of machines, all of the fitting to the gauge would have been done by filing.

After filing to shape, the Academicians state: *The lockplate is fixed firmly on a sufficiently thick jig to direct the punch perpendicularly. In this jig are pierced all these holes exactly in the position, and at the respective distances from each other, as in the finished lockplate. We start by marking them with a punch on the new lockplate: this operation done, we finish drilling them with another slightly conical drill.*

'Punching' seems a somewhat brutal treatment of something upon which so much care has been exercised up to this point. They then tell us that in the next operation the holes *are then made cylindrical, by means of a tool called a faux-foret* [false-drill], *used with a conductor* [pilot? guide?], *so that these holes are perpendicular to both* [front and back] *surfaces: finally, with a reamer, they are given the determined size, & afterwards the appropriate ones are tapped.*

Exactly what form a *faux-foret* (false-drill) took has thus far proved elusive but, with a plain extension to act as a pilot to guide it in the hole, it could be similar to a 'counterbore' or a reamer as shown in Fig. 5.15.

This description is very different from the way the operation was performed at the Atelier de Perfectionnement in 1794-96, judging by an illustration of the *Machine a percer les trous de la platine*, Fig. 5.16.

Here it would appear that the same hole was drilled repeatedly in a succession of lockplates, followed by the jig being re-adjusted or a series of jigs being used for

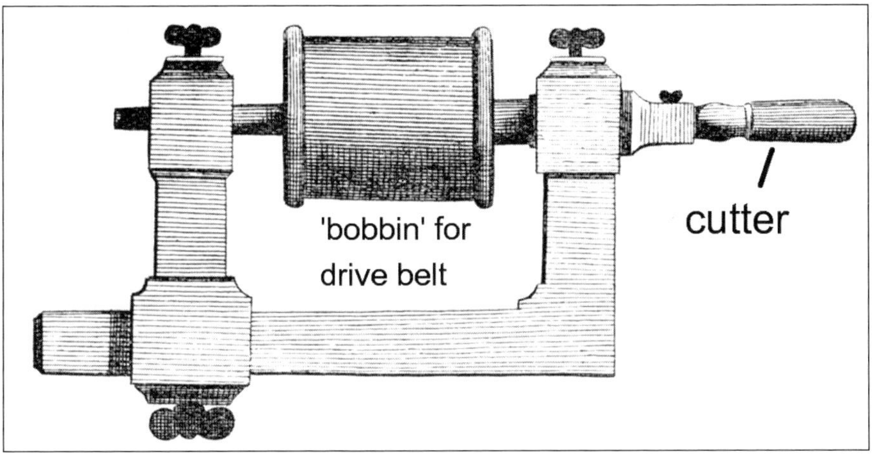

Fig. 5.14. 'milling machine' using a 'rotary file' to finish the inside surface of the hollow in the pan. (Diderot, 1751, Pl. 4)

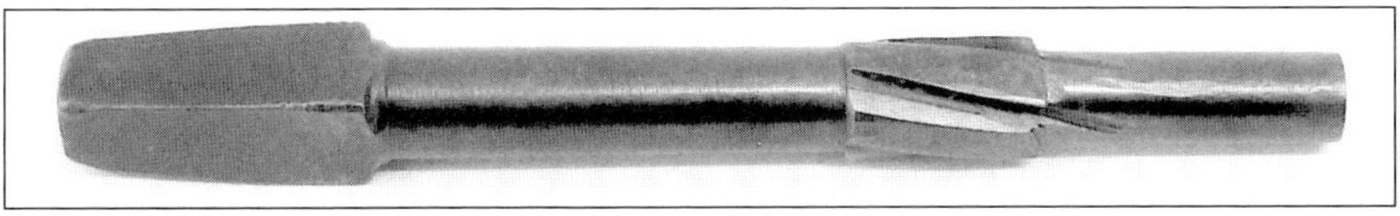

Fig. 5.15. A counterbore/reamer – 'faux-foret' – false-drill? (Author's collection)

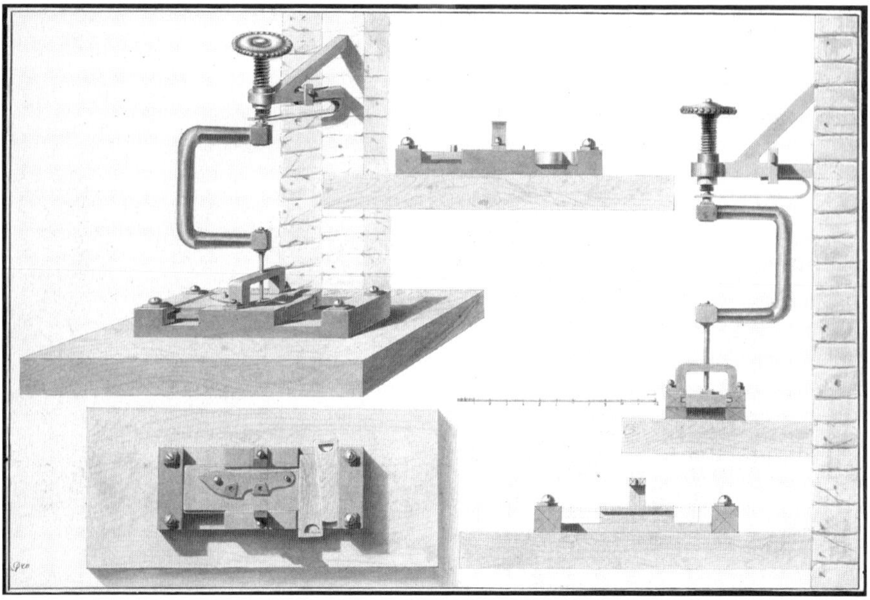

Fig. 5.16. A Machine for drilling holes on lockplates of muskets. (drawing by Dromard, late 18th century. Lnv. 13571-.418, © Musée des arts et métiers-Cnam, Paris/photo Dephti-Ouest)

the next series of holes, and so on, since the guide bridging the lockplate has only one hole and one position. The re-setting of the jig must have presented difficulties since it was held in position by 'folding keys', or wedges, although it may have been possible to re-position the holding device if the table on which it was fixed had a series of accurately positioned holes for the securing screws. And, with the brace only being supported at its upper extremity, it has to be assumed that the drill itself was a close fit in the guide to ensure both accurate positioning of the holes and that they were perpendicular.

Closer examination of Fig. 5.17 reveals what appear to be shadows, suggesting the lockplate was orientated by dowels passing through the tumbler and the front screw, or *side-nail,* holes, Fig. 5.18, although this is not stated in Cotty's description.

If this were the case, then these two holes would need to be formed accurately in size and juxtaposition to enable such a method to be used and Cotty's comments[23] are suggestive of this: *On removal from the stamp it is placed in a machine, the holes of which serve as guides to the different punches with which all the holes are pierced, except those of the rear and front which are pierced by the drill . . . It should be observed that the die in which the protuberances flanking the pan opening are cut, does not have a depth equal to the thickness of these parts, in order not to damage the lockplate by deforming the iron in these places; the rest of the operation is done with a chisel.* However, it would again appear that not all the holes were drilled but his reference to a 'chisel', which might be interpreted as 'graver', is obscure.

It is interesting to note that in punching the holes in the 'protuberances' – the *trapezoidal eminences* – they rested firmly upon a solid support. He implies that another die or jig was used for this, Fig. 5.19, which would be necessary because, if the same jig was used for punching these holes as for punching the other holes in the lockplate, there was a risk of deforming the lockplate.

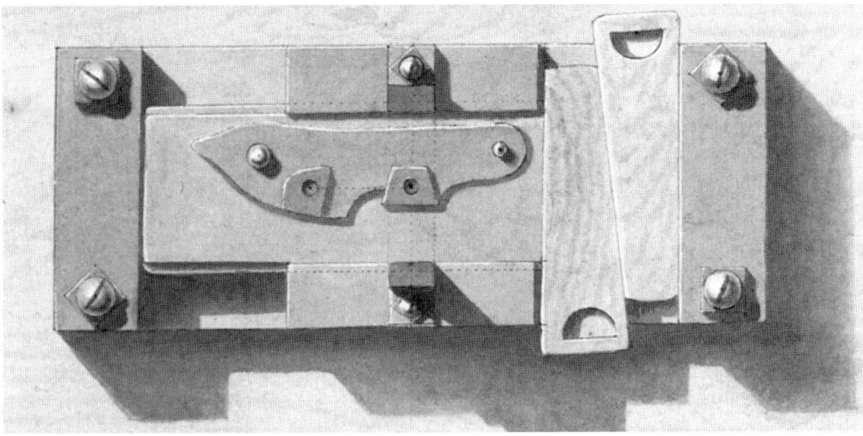

Fig. 5.17. Detail of the drilling jig in Fig. 5.16 with the guide-bridge omitted – note the lock is placed outer-face down. (drawing by Dromard, late 18th century. Lnv. 13571-.418, © Musée des arts et métiers-Cnam, Paris/photo Dephti-Ouest)

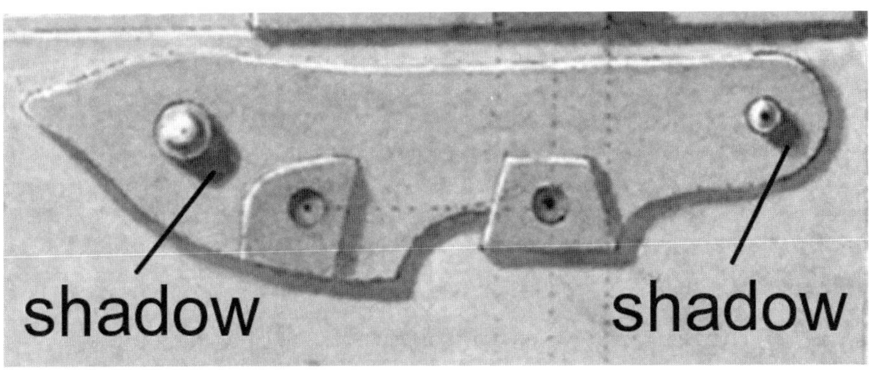

Fig. 5.18. Showing what appear to be shadows from projecting dowels. (drawing by Dromard, late 18th century. Lnv. 13571-.418, © Musée des arts et métiers-Cnam, Paris/photo Dephti-Ouest)

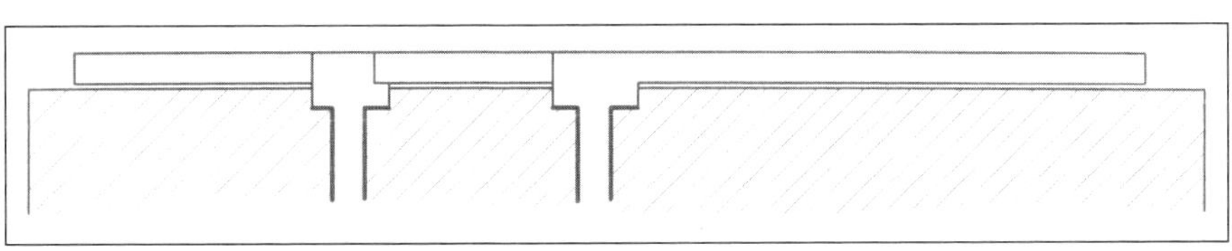

Fig. 5.19. Interpretation of Cotty's statement that in punching the holes on the 'protuberances', the recesses in the die used were shallower than the thickness of those 'protuberances'. (© P. Smithurst 2018)

In what seems a self-contradiction, Cotty also states: *The way of drilling the holes is more accurate than the ordinary one, where the worker applies a template to the lockplate, for example, to mark the holes on it with the drill bit and then drill them. Good workmen can, it is true, work by this last method with a great deal of precision.*[24]

The Tumbler (Noix)

The tumbler [*noix* = nut in French terminology] was regarded by the French Artillery Officers in charge of arms manufacture as 'the brains of the lock'.[25] This is an apt description since the tumbler was central to, and controlled, the lock's operation. The tumbler carried the cock, was acted on by the mainspring and engaged with the sear which enabled the gun to be cocked and then fired. The tumbler, therefore, is a complex object requiring accuracy in its completion. It was of paramount importance that the square on the large diameter shank, to which the cock was fitted, was both accurately formed, not only in its geometry but in the relationship of that geometry to the notches which had to be formed. These notches engaged the sear which held the tumbler, and therefore the cock, in its two positions – the half-cock 'safe' position and the full-cock 'firing' position.

As the authors of the Report note: *In fact, it will be easy to conceive, if we reflect on it, how difficult it is, for example, to make a tumbler which, substituted for another in a lock, connects so perfectly with the sear, that the rest* [position] *of the cock be as well assured as before, and that the detent be made precisely as it ought to be done; finally, that the cock is situated on this new tumbler exactly as it was on the old, so that it traverses the same space, and falls on the frizzen in the same place as before, to be certain of sparking.*

Fig. 5.20, shows a M.1777 tumbler alongside the official French drawing.

le Roy et. al. in their report[26] then state: *The tumbler having been prepared, with its large rounded pivot, it is reduced to its true thickness by milling its two surfaces.*

Fig. 5.20. Official drawing of the tumbler and its gauge of the M.1777 lock, c. 1794-6. (Alder, personal communication) alongside a M.1777 musket tumbler.

This same process is also noted by Cotty:[27] *It is made to correct thickness with a tumbler mill, as with other locks; its whole outline is formed in a stamp with a die, the depth of which is less than the thickness of the tumbler body, in order to prevent the lower edge from becoming uneven; the remaining iron* [presumably the 'fin' squeezed out between the die faces] *is removed by a chisel.*

Obviously, the body had to be thicker than required for it to be 'faced' to correct thickness in the tumbler mill!

Diderot illustrates a tool for machining to the correct thickness, Fig. 5.21.

Its name is given as '*Calibre double avec la fraise ses parties d'assemblage*' [jig with double milling cutter for parts of the assembly]. It is different from the '*calibre double*' of Alder which he suggests was used to file the square on the large tumbler shank to receive the cock with its matching square hole,[28] but it may have been a more generic term. In Diderot's reference to 'milling', the 'milling cutters' can be seen to be file-cut plates, but its operation is apparent from the illustration although the purpose of all the holes in the top plate is uncertain. The thicker of the two shafts of the tumbler was entered in a hole in the bottom 'mill', *5*, the small arbour was entered in the hole in the top 'mill', *3*, and the two 'mills' brought together by the two wing nuts, *1* and *2*, while rotating the tumbler relative to the mill.

This causes the corners of the holes in the file faces to both gradually 'round' the two arbours and eventually bring them into contact with the faces of the tumbler body. With continued rotation and equal tightening of the wing nuts, the tumbler body was 'milled' to the correct thickness with both faces parallel. Exactly how parallelism was accomplished is not stated, though a simple expedient would have been to have collars of equal thickness to the finished tumbler on the two clamping screw shanks to act as 'stops', ensuring both correct thickness *and* parallelism. Diderot's illustration is slightly earlier in date than Blanc's work on the 1777

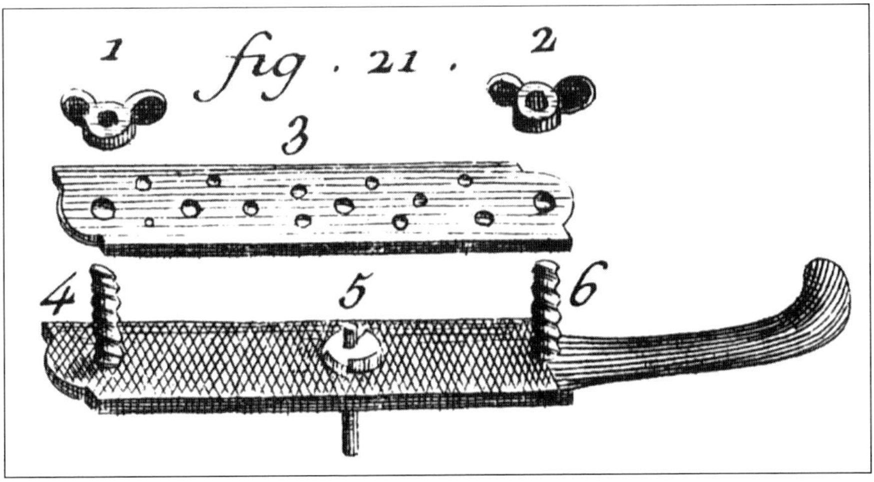

Fig. 5.21. A *calibre double* - a 'tumbler mill' illustrated in Diderot's *Encyclopédie*, consisting basically of two files which act on both faces of the tumbler body at the same time when the tumbler is rotated. (Diderot, 1751, pl. IV)

lock and possibly Blanc's 'tumbler mill' was a little more refined. Alder states that *after 1763* [the date of Blanc's promotion to controller of the arms factory in Saint-Etienne] *the tumbler's large pivot/arbour was rounded with a tool which had the effect of a lathe* and that this tool *was almost certainly Blanc's hollow milling machine, the first of its kind.*[29] If he is just referring to a tumbler mill, then it has already been shown by Diderot that a similar device existed in 1751, and it is equally likely that such machines were also in use in other gunmaking centres.

Of course, Blanc's hollow end mill achieved more than simply rounding the pivots; it also machined the faces of the body, and a machine of this nature, utilising 'hollow end mills', was in use at Tula a few years later and was certainly used in English practice. After being machined, le Roy et. al. note that the tumbler: *is mounted between hardened jigs exactly of the same contour which the tumbler must have and is filed on it; But we do not limit it; After having thus filed it, it is passed through a die made expressly, so that the notches will be perfectly formed.*

Cotty also notes that in a similar fashion: *The square on the arbour is formed in another die; it is pressed, when filed, into a square die of sufficient size to give it its exact dimensions, then in another square die so it can be cut it to its length, and finally in a third jig to drill and tap the hole of the screw.*[30]

Cotty later comments that a narrow and shallow raised portion is formed around the large shaft to reduce friction[31] but this is not shown on the official drawing, nor is it apparent on a tumbler from a M.1777 lock, Fig. 5.20. However, an 'anti-friction' collar is formed on various lock components of the Pattern 1853 Enfield rifle.

For the manufacture of the other lock components, Cotty is the major, if not the only, source of information from this period, along with Dale's account of hardening them.

They were first hand-forged to approximate shape and size and: *When there is a certain quantity of pieces thus forged, they are made red-hot in a charcoal fire; then they are placed, one after the other, in a die which, of course, is changed as soon as the pieces of the same species are stamped. All these dies, made of iron and bound with strong bands of steel* and the pieces, well heated, are successively subjugated with solidity on the anvil, where the fall of the hammer gives them the shape of their die. If, as often happens, pieces have been forged too large in size, they form fins that are removed by shears; after which they are made red hot again, and returned to the dies, of which the two parts which compose them must come together.*

In regard to the material used for dies, by this time 'crucible steel', or 'cast steel' as it was often referred to, developed by Benjamin Huntsman in Sheffield about 1742, had been in use for around sixty years. Its advantage over the only previously commercially available steel, 'blister steel', was that it had been produced in a molten state and was thoroughly homogeneous and therefore lacked the slag inclusions of blister steel which might cause areas of weakness in a highly stressed component such as a die. By the 1780's, Huntsman's son, William, was

* *one makes the body of these stamps out of iron, because if they were entirely made of steel, they would be too fragile.*[32]

supplying Matthew Boulton with steel for making coining dies.[33] It is also known that Huntsman's steel was being supplied to France around this time, albeit for cutlery, so obviously some were aware of the existence of a steel of unsurpassed quality and even a Swiss commentator noted in 1778: *The cast steel of England is, without contradiction, the most beautiful steel in commerce.*[34]

The Frizzen *(Batterie)*

Unfortunately, the original drawing of the frizzen is too pale to reproduce satisfactorily so a photograph of the actual item is provided.

It was forged in a die and: *Its entire outline is filed against a template [Although the pieces coming out of the die have almost always its internal imprint, however, for more regularity, we file the contours on the templates] which also serves as a jig and guide for drilling its screw hole: in general, the thickness of the jigs is sufficient to serve as a guide.*[35]

The worker had to use the frizzen of the model to make the slope on the frizzen body the same and to adjust the plane of the body and the curvature of

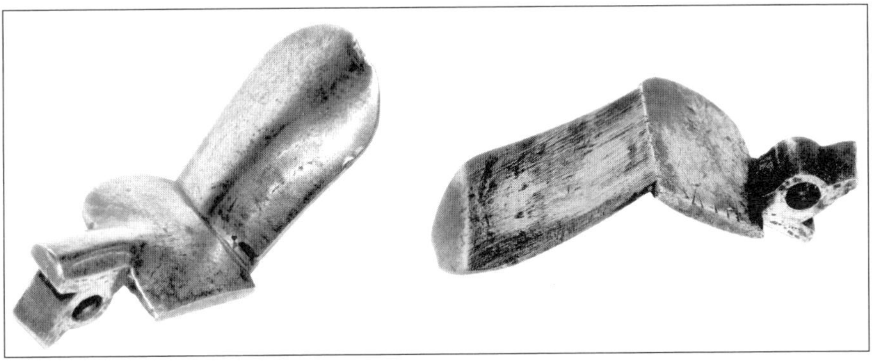

Fig. 5.22. A frizzen *(batterie)* from a 1777 lock. (Author's collection)

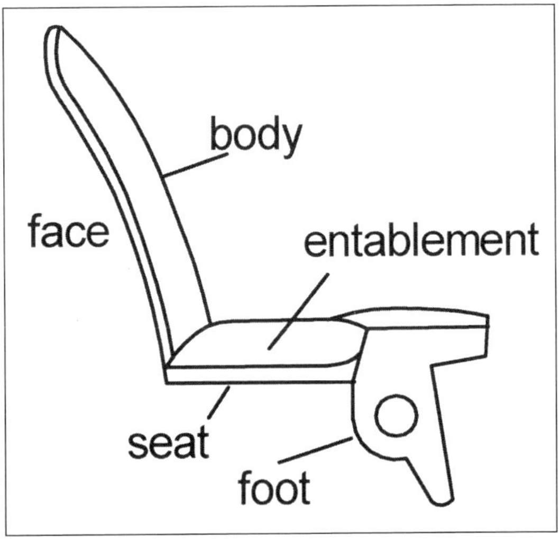

Fig. 5.23. Terminology of the frizzen *(batterie)* used by Cotty. [the term 'entablement' is the same in French and English and means 'supporting platform'] (© P. Smithurst)

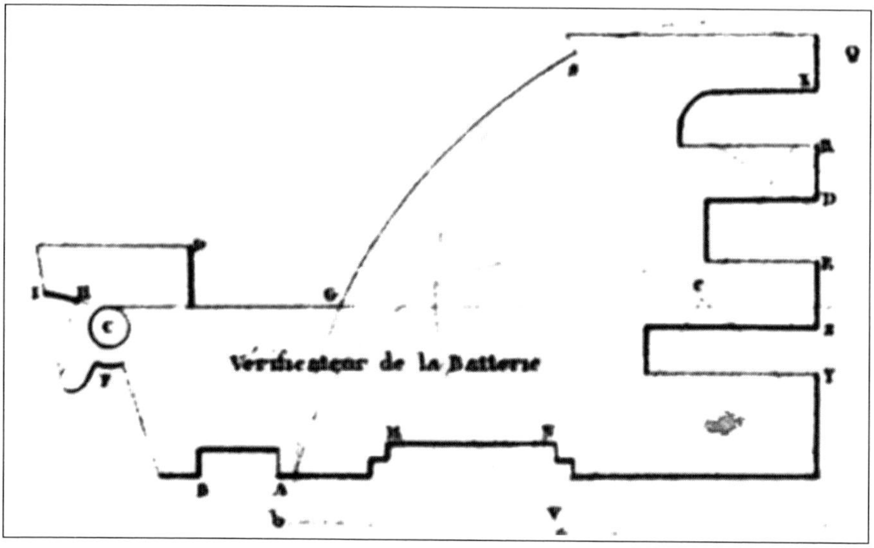

Fig. 5.24. Official drawing of the *frizzen* gauge of the M. 1777 lock, c. 1794-6. (Alder, personal communication)

the face. It had to be ensured that the face of the frizzen over its whole length was square with the lockplate. The flat seat was placed on the pan and a square applied over the length of the curved face of the 'body'. This square used by the lockmaker had a recess to accommodate the pan. To ensure the frizzen produced sparks when struck by the flint: *The face must be covered with a 0.23 [1 l] thick steel sheet in a musket and 0.19 [10 p] in a carbine.*[36] This implies a separate piece of steel which is at odds with the case hardening process described by Dale later. If correct, it might provide a means of ascertaining a date or place of manufacture of a musket.

The Cock *(Chien)*

This was produced in a similar way to the frizzen:

> *Its outline is filed on a template; the square and the heart are removed by stamping/punching in dies (the heart must be roughly formed at the forge, and the center of the location of the square, drilled in the drill, before passing to the stamp, to facilitate the operation of the punch), the hole in the lower jaw is made by the drill, in one jig; that of the upper jaw, whose contour is formed on a jig, is made in the same way.*[37]

The holes referred to as the 'square' and the 'heart' are apparent from the drawing below, Fig. 5.25. The notations on both drawings are largely illegible, even at larger scale, but from the few that are not, it can be deduced that they relate the dimensions

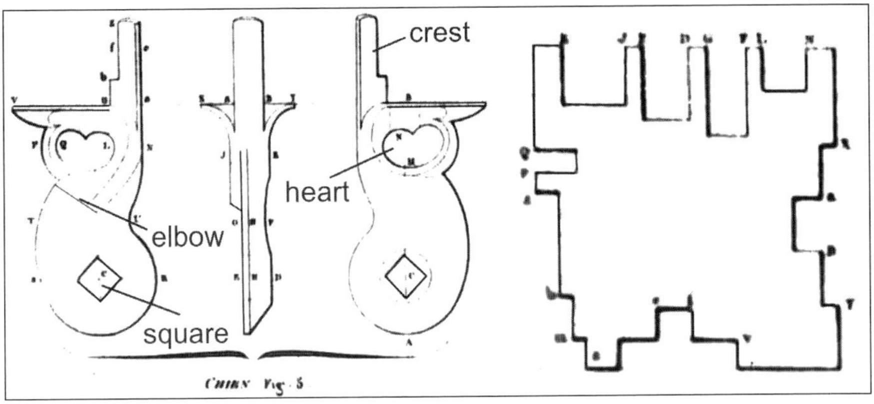

Fig. 5.25. Official drawing of the M. 1777 cock and its gauge [with annotations added]. (Alder, personal communication)

of features on the cock to the parts of the gauge where they were checked. Thus, the four at the top correspond with thicknesses of various parts of the cock body.

Scrutiny of the set of gauges by Blanc shown earlier reveals what may be an additional pair of gauges for the profile of the cock, Fig. 5.26, but how they were applied to such three-dimensionally contoured surfaces is not immediately apparent.

Cotty notes that several gunmakers thought that some cocks, more particularly those that were die-stamped, very often broke at the throat and it was suggested it would be advisable that this portion was formed by bending at the forge. It was argued that: *this would be stronger by making the fibres of the iron follow the curvature and which, even if not perfectly coinciding with the shape of the throat, would at least prevent it from breaking. However, this process would have required iron of excellent quality* [which would have been expected anyway in making lock components] *and many precautions on the part of the worker.*[38] Given the nature of the 1777 cock, this comment seems superfluous since the throat already has additional support from the 'strut' forming the front of the 'heart' and is more applicable to the graceful 'swan-necked' cocks to be found on earlier military and on sporting guns.

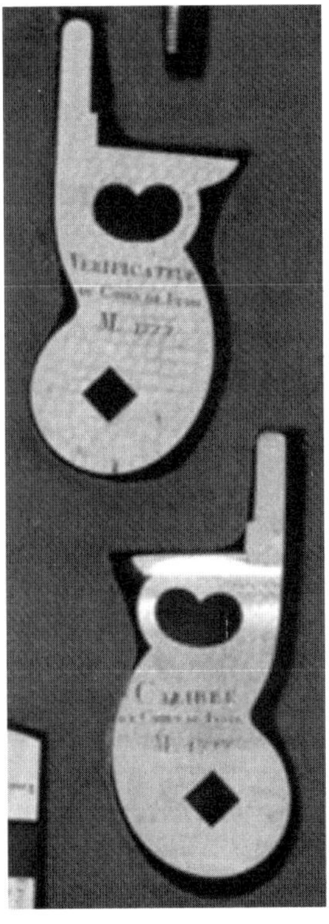

Fig. 5.26. Detail from the set of Blanc's gauges showing what appear to be profile gauges for the 1777 cock. (Inv. No. 2013.0.11, Photo (C) Paris - Musée de l'Armée, Dist. RMN-Grand Palais - © Emilie Cambier)

The Bridle *(Bride de noix)*

The bridle served to support and partially encase the tumbler and the sear.

Its holes were drilled on a machine, and its outline filed by means of filing jig. It was noted that it was of the utmost importance: *that the holes in this corresponded exactly with those of lockplates and, in general, if the templates, jigs, etc., did not relate perfectly to each other, its manufacture is very difficult.*[39]

The Sear *(Gâchette)*

Cotty indicates that the hole for its axis pin was drilled in a jig and its nose, having a fanciful resemblance to a bird's head and beak, which engages with the tumbler notches, is filed against a jig.

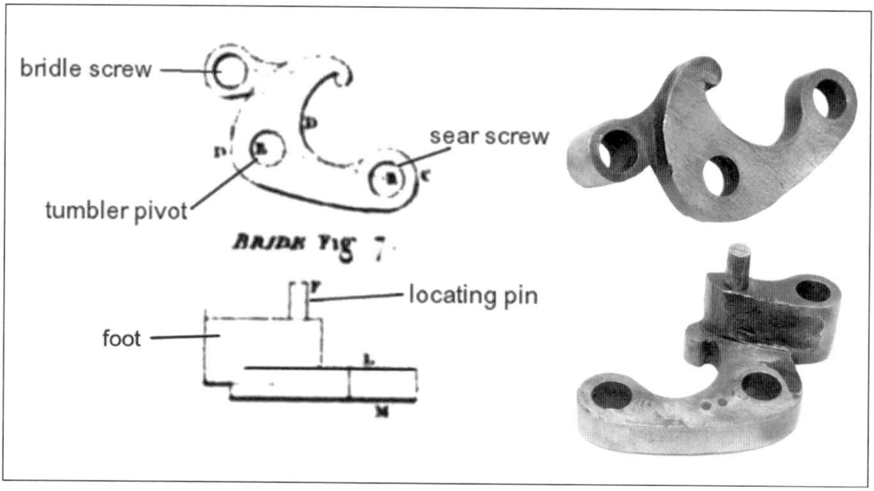

Fig. 5.27. Annotated official drawing of the M. 1777 bridle. (Alder, personal communication) alongside a M.1777 musket bridle. (Author's collection)

Fig. 5.28. Official drawing of the sear. (Alder, personal communication) alongside a M.1777 musket sear. (Author's collection)

The Springs *(Ressorts)*

Most flintlock arms have three springs: mainspring (*grand ressort*), sear spring (*ressort de gâchette*) and the frizzen spring (*ressort de batterie*), Figs. 5.29 and 5.30.

The 'locating pins' of the mainspring and frizzen spring are clearly shown and Cotty states that these *are rounded by means of a milling cutter, which at the same time makes the underside square.*[40]

It should be observed that Cotty makes no mention of a 'nib' on the sear spring to engage with a slot on the inner face of the lock plate and which prevents it rotating about its fixing screw. Such a feature was important and common to every flint, and later percussion, gun lock, both military and sporting.

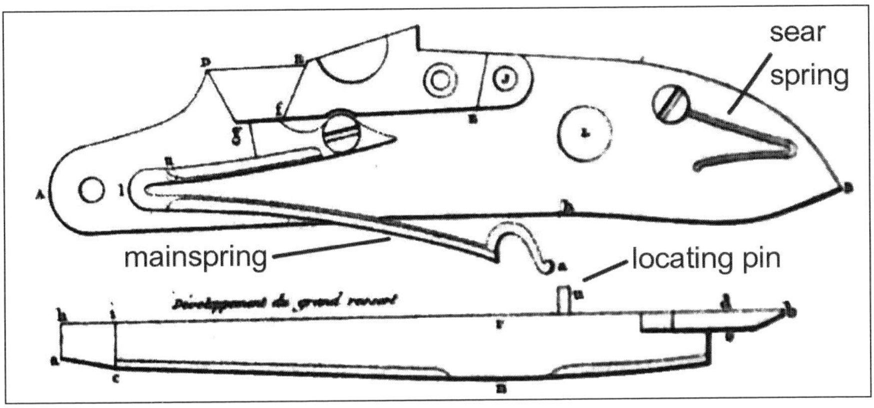

Fig. 5.29. Annotated official drawing of the mainspring and sear spring in place. At bottom is a top-view 'development' of the mainspring showing it straightened from its 'V' form. (Alder, personal communication)

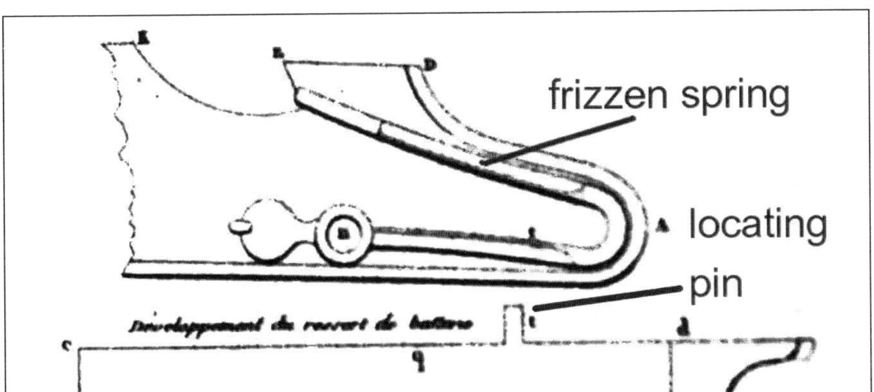

Fig. 5.30. Annotated official drawing of the frizzen spring in place. At bottom is an underside view 'development' of the spring showing it straightened from its 'V' form. (Alder, personal communication)

The Screws (*vis*)

These were made in the same manner as the cock screw but: *One trick to abbreviate the work is by slitting at the same time a certain quantity of screw heads.*[41] Dale notes that the large lock screws, or 'side nails', were threaded and finished and their holes in the lockplate tapped as part of the setting-up process.

Hardening and tempering the lock components

Cotty has nothing to say about this important aspect since it fell to the assemblers of the musket, "*l'équipeur–monteur*" (setters-up in English terminology) to perform this work, but it is extensively covered by Dale's descriptions of stocking etc. published in 1810 and his narrative will be drawn upon.

It has already been mentioned that Blanc initially had problems with the warping of components during quenching and noted that in English gunmaking circles the 'art' had been mastered. He eventually overcame these problems and hardened items successfully. It may seem odd that, like freshly felled wood drying out, metals can warp when thrust into water while red hot. It is due to the parts of an asymmetrical shape losing heat at different rates. One of the best examples witnessed was in the days of file making in Sheffield some decades ago. A 'half round' file will lose heat faster from the round face than from the flat face because the former has a larger surface area. If, after cutting the teeth, a straight 'half round' file is heated to red heat and plunged into water, it will bend towards the curved face. To overcome this, legend has it that file hardeners used to bend the soft file in the opposite direction, so it straightened on quenching!

Nearly all lock components are asymmetrical, but they are also small so any tendency to warp will be much smaller. It is assumed that since Dale does not mention this problem it had been solved.

The processes which must be carried out according to the different nature of these two substances (iron and steel), generally result in two kinds of hardening. That which is applied to iron is known as case hardening, or 'cementation', and that applied to steel is simply quenching when red hot.

Case-Hardening

The objective of case hardening is to give pieces made from wrought iron a sufficiently thick 'skin' of steel so that when quenched at red heat this steel skin is hardened. Treated in this way, parts can better resist wear from continual rubbing, especially where reduced to small dimensions such as the tip of the sear, notches on the tumbler, the nose of the tumbler and claw of the mainspring rubbing together, the threads of screws, the base of the frizzen rubbing against a steel spring etc. Also, if unhardened, the heads and slots of screws are prone to deform through the use of the screwdriver. Hardening also gives iron the property of receiving a better polish and more resistance to corrosion, an essential attribute for weapons which are often exposed to the elements.

Case hardening in which the surface of the iron is converted into steel, was traditionally called known as 'cementation' and workers in the past referred to the products as being 'carburized'. At red heat, carbon from the crushed and sifted 'hard charcoal' used in case hardening, penetrated iron to a small depth and chemically united with it to form a skin of steel.

As Dale described the operation,[42] which has changed little, a layer of charcoal was placed at the bottom of a circular cast iron container, and on it were placed the first layer of pieces to be hardened. They were then covered with a layer of charcoal and a second layer of pieces added, and so on, until the pot was full. The container was then sealed with a thick layer of clay to exclude air (oxygen). It is very important to exclude air, otherwise the charcoal would burn away, and the pieces would not become hardened. The clay for the cover was also free from combustible material.

It was also very important that the pieces did not touch each other or the walls of the pot because if the charcoal did not cover all their surfaces, those portions would not be carburized.

As these different parts needed to be case-hardened to different degrees, a distinction was made between the large and small items during hardening. The large items were the lockplate, frizzens and cocks, and the rest were the small items. A further division was made within the 'large items' since the frizzens had to receive a deeper case hardening than the lockplates and the cocks. Hardening was thus broken down into three groups: frizzens; lockplate and cocks; small pieces.

The frizzens were hardened in batches of fifty, occupying three layers and laid on their sides to fit together and occupy the least space possible. The lockplates and cocks, twenty-five of each, also formed three layers, the lockplates being placed on edge and the cocks filling the empty spaces between the lockplates and the walls of the pot.

The small pieces were placed in four layers, each package containing enough for twenty-five muskets, but smaller numbers were hardened when necessary, so as not to mix those of two workshops. Hardening the pieces for fifty muskets therefore comprised five batches.

Dale also notes the setter-up should remove sharp edges from the lockplate on a grindstone before and after hardening. This had to be done by an experienced and intelligent workman before the lockplate could be polished.

Inspection of lock components

The lock components were inspected after hardening and while the hardening process fell to the setter-up and although Dale offers some commentary on the inspection process after case hardening,[43] Cotty provides extensive details.[44] Gauges are not mentioned despite their existence and their appearance on official drawings. Cotty's details have been included in their entirety since they are not only important in their own right, but also provide the functional requirements of relevant features of each component. Where it is felt necessary, particularly for non-firearms readers, additional comments have been included in square brackets. It should also be noted that at the time Cotty was writing, he gave decimal measurements first, in centimetres, followed by the old measures in brackets: 1 point (p) = 0.188 mm, 1 ligne (l) = 12 points = 2.256 mm

The Lockplate (*corps de Platine*)

The mechanism of the lock is such that, if one of these pieces is incorrect in size or position, it results in more or less serious inconveniences. Its effect depends mainly on the relative forces of its springs, and the respective positions of the pieces that will be examined successively.

The inner face of the lockplate must be perfectly flat, as well as the outer face, excepting the tail, which is rounded, so that all the parts that must fit are truly square to it. All holes must be drilled perpendicularly, cylindrical, and well threaded. If they were not located as they should be, and if the lockplate did not have the prescribed thickness, the resulting lock would not be fit for service.

The length of the lockplate, with all the military locks, must be five times the thickness of the barrel at the breech; its widest width, near the rear of the pan, must be one fifth of its length, and its thickness 0.51 (2 l. 3 p) for the infantry musket and 0.45 (2 l) for the carbine; that of the boss [trapezoidal eminence] *for the frizzen screw, including that of the body, from 1.13 (5 l) to the first of these models and from 1.01 (4 l 6 p) to the second; finally, that of the boss for the lock fixing screw of 1.07 (4 l 9 p), also in the first musket and of 0.96 (4 l 3 p) in the carbine.*

The Pan (*Bassinet*)

The recess made in the lockplate to receive the pan must be long enough and deep enough so that the distance between the bottom of the recess and the bottom of the pan [where the powder is placed] *has to be 0.45 (2 l) on the musket and 0.40 (1 l 9 p) on the carbine. The pan must seat well on the lockplate; its milling* [of the hollow] *must be very straight, and the inclined plane of its lateral edges well squared with that of the lockplate; its total length in the direction of the milled hollow must be, on all the military lockplates, equal to the greatest width of the lockplate.*

The Frizzen (*Batterie*)

If the frizzen is too sloped, the cock may be at the end of its course before the pan is uncovered; and as the flint, during the striking, would not rub on a sufficient length, it would result in few or no sparks. If it were not sufficiently sloped, the flint would strike it too low, and the same inconveniences would still occur. The face must be squared on the lockplate, that is to say, it must not lean on any side: otherwise, the flint could only scrape it at one point. The entablature should form a right angle with the plane of the buttress of the lockplate (the pan being uncovered); but if it were not exactly right, it would be better to be obtuse than acute; in the first case, the frizzen moves its fire away from the center of the pan; in the second it is less easy to initiate. Finally, if the frizzens do not fit closely on the pan, the primer can be lost.

The worker must work to the frizzen of the model to have a slope on the frizzen the same and to adjust the plane of the body and the curvature of the face.

To make sure that the face of the frizzen is square with the length of the lockplate, the entablature is placed on the pan, then one of the sides of the square is applied to the length of the body, and we see if it sits on every point of its face on the other side. This square used by the lockmaker has a recess to accommodate the pan.

The height of the face, measured from the angle it makes with the seat, is equal to the distance from the center of the hole in the shank of the tumbler to the center of the inner part of the trough of the pan; its widest width is very nearly one-sixth of the total length of the lockplate, and its base is 0.45 [2 l] less than the musket and carbine. The width of the seat is equal to the length of the pan in these two models.

The foot of the frizzen must be well drilled in its centre and should be straight, rounded over and sit squarely on the spring.

The Cock (*Chien*)

So that the cock in the 'fired' position is in front of the frizzen, it is necessary that the center of the hole of its square is on the axis of the hole of the lower jaw; [Fig. 5.31] *for if it is in front of this line, the lower jaw moves away from the frizzen, and if it is behind, it approaches too closely: in either case, the sparks do not enter the middle of the pan. The same inconvenience will still occur if the height of the cock, taken from the center of its square to the lower jaw, is not equal to that of the angle of the entablature of the frizzen at the upper part of its face.*

The top jaw must be drilled so that with the screw in place, it is inclined to the bottom jaw, so that it can pinch the flint at its end, and the heel of the top jaw must be pressed against the crest of the cock [this is the 'column' projecting from the bottom jaw]*: these two jaws must be equal in length and width. The top jaw screw must be perpendicular to the lower jaw, and parallel to the crest in all its length, otherwise it could break.*

It is essential, first, that the thickness of the elbow is from 1.01 [4 l 6 p] on the musket and 0.90 [4 l] on the carbine, so that it does not escape from its support;

[It needs to be pointed out here that Cotty seems to be in error if we take the feature he refers to being the 'step' on the inner face of the cock which, when in the 'fired' position rests upon the top edge of the lockplate. While we might equate 'elbow' with 'shoulder', a 'thickness' of 1.01 cm is not credible, but might apply to its length, Fig. 5.32. On the other hand, Dale notes ". . . *if the shoulder . . . were too narrow, the cock could fall too far and damage itself or the lockplate*".[45] Measurement on a 1777 cock shows this shoulder or step to be 3.75 mm.]

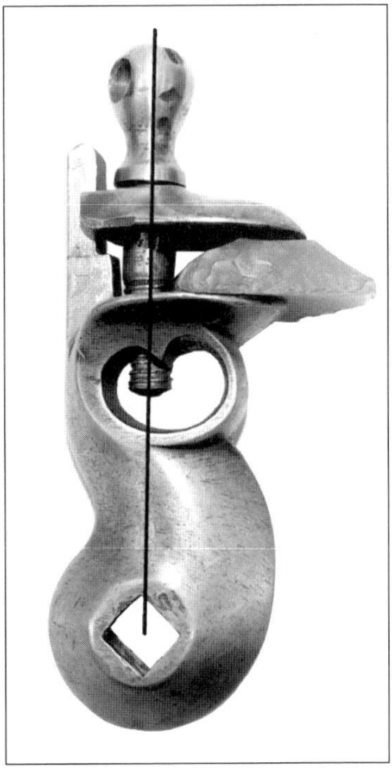

Fig. 5.31. 1777 musket cock with centre line drawn. (Author's collection 2022)

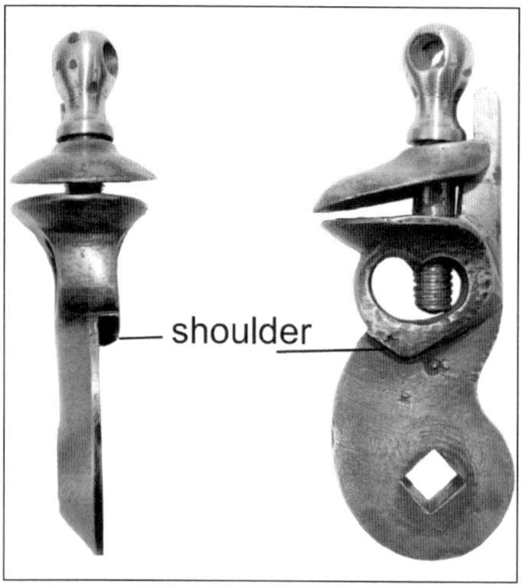

Fig. 5.32. 1777 musket cock showing the shoulder. (Author's collection)

Second; that the square is very uniform, that it bears in all the points of its interior surface, and that the thickness of the cock at this point is from 0.74 [3 l 3 p] on the musket and 0.63 [2 l 9 p] to the carbine, and that it is held securely, and third, that the thickness of the lower jaw where the screw passes, be 0.68 [3 l] on the musket, and 0.62 [2 l 9 p] on the carbine, and so that the screw is sufficiently restrained there it must fit correctly in the upper jaw, and is not threaded at this part.

The Tumbler (*Noix*)

It is very important that the stem of the tumbler is a perfect fit in the hole of the lockplate, that the round portion must extend a little beyond the outer surface to avoid the rubbing of the cock against the lockplate; for if it were not so, the mainspring, by the pressure it continually makes on the claw of the tumbler, would draw it towards the lower part, and cause friction; the cock, following the movement of the tumbler, would also rub in the upper part on the lockplate, so that the force of the mainspring would be considerably diminished. Lastly, the stem bearing only two opposite points, one below and the other above, would become more and more loose in this hole and the friction of the tumbler and the cock would soon increase to such an extent that the frizzen would no longer be struck forcibly enough to work.

If the claw is too deep, the mainspring loses its strength because it cannot be brought to the necessary degree of tension: it must be raised for that to happen to the height of the axis of the tumbler. If, on the contrary, the claw is too shallow, it will drive the long arm of the mainspring back on itself and render the lock difficult to operate, resulting in breakage of the mainspring.

If the tumbler does not have enough circumference, the scallop will come too close to the axis of its stem, so that it will bind on the sear, making the lock difficult to operate. The re-entrant parts of the notches for the 'firing' and 'safe' positions must be on the same arc. It must be the same for the protruding part of the notch of the 'safe' position for if it exceeded this arc, the sear nose, escaping from that of the 'firing' notch, would meet it, and one or the other would break, or the cock would be held back in its motion. It is to avoid this serious defect, which the lock maker is called to meet, that he disposes this notch so that its salient part is, compared to that of the 'firing' notch, rather a little inside the arc than outside.

The thickness of the tumbler should be 0.68 [3 l], in the musket, and 0.62 [2 l 9 p] in the carbine. The diameter of the stem must be 1.13 [5 l] in the first of these models and 1.01 [4 l 6 p] in the second; the square is inscribed inside the circle of this stem, and the thickness of the cock determines its length. The hole of the tumbler screw (incorrectly called the cock nail) must be drilled in the middle of the square. So that the claw is not too long, it is necessary that by turning the tumbler in its hole (it must be free for that), it can be checked that this claw does not project beyond the edge of the lockplate.

On the part of the tumbler facing the lockplate, a small collar of about 0.02 [1 p] thickness is formed in all military locks, to prevent it from rubbing its entire surface on the lockplate. [This collar has not been found on the few muskets examined.]

The Bridle (*Bride de noix*)

If the outside of the flange, at the height of the hole of the pivot of the tumbler, exceeds the circumference of this tumbler, the great spring, whose claw would then bear against it, could not fulfil its object. If the depth of the bridle at the same place is too low, it will break.

If the hole that receives the [small] *pin of the tumbler is badly drilled, as well as those of the foot fixing screw, the pivot of the flange and the sear screw, the flange that must support the tumbler vertically (the lock being fitted to the musket) will push it in the opposite direction and will disturb the operation of the lock by considerable friction. It is necessary that the thickness of this piece be at least 0.45 [2 l] on the musket, and 0.42 [1 l 10 p] on the carbine.*

The Sear (*Gâchette*)

The sear must have the thickness of the tumbler, and one reserves right around the hole a collar of the same size as on the tumbler and for the same purpose of reducing friction. The length of the 'nose' is equal to the distance of the screw to the edge of the hole of the shaft of the tumbler.

The sear nose, to be able to move against the force of the mainspring, when the cock is in 'full-cock' position, must be strong and must mesh well with the notches of the 'cocked' and 'safe' positions of the tumbler, otherwise, the cock may fall down (1) Its tip must be rounded, for without this, it could be caught at one or the other of the projecting parts of the notches, if the soldier had not taken the precaution of checking whether it had engaged well.

(1. in the exercises, the 'full-cock' is sometimes left for a long time, without thinking of the consequences that may result to the frail sear nose)

Springs (*Ressorts*)

All springs must be well bent; it is from this quality and from that of the steel that their elasticity depends: excess thickness deprives them of their flexibility and force.

The small branch of the mainspring must fit well on the lockplate, and the great branch must have an equal clearance of the lockplate in all its length; if it experiences friction, the lock is stiff in its movements. The shoulder of the short branch must rest against the rampart of the frizzen [trapezoidal eminence, the thickened part of the lockplate in front of the pan]*: it facilitates removal of the* *screw, the adjustment of the spring on the lockplate, and relieves stress on the pivot. If this rampart were not there, the spring would be able to press against the pan, which would cause it to rise and disturb the adjustment of the frizzen; moreover, the pivot and the screw would not last long from the pressure of this spring. Neither is the large branch free, its width taken from the outer part of the claw to the lockplate must be a little from 1.13 to 1.55 cm.*

The claw of the great spring must not be too long, nor too deep; the first defect can only take place at the expense of the long branch which, being strained, loses its elasticity; the second would be its being trapped against the tumbler claw before the cock reached its limit, and to do that, wood has to be removed near the lock screw, which would be detrimental to the service of the lock. If, on the contrary, this claw was too short, it must be reduced to that of the tumbler, so that it would be difficult to cock the gun. But, among other things which might result from this operation, the extremity of the claw of the spring would be at the extremity of that of the tumbler and would be at risk of disengaging from the tumbler when the lower part of the latter descended, as there is only about 0.11 cm spare above the edge of the lockplate.

It will be observed finally: first, that the end of the claw of the mainspring must be exactly on the prolongation of the outer plane of great branch; second, that the distance between the end of the claw of the tumbler and the bottom of that of the mainspring, when the cock being drawn back, must be approximately 0.07 cm for the cock to have the necessary movement. (The term "surbande" is the distance the cock can still take when fully cocked.)

The frizzen spring must have very little play in its great branch, and the little branch must be well seated; the bend should be placed at the start of the front bevel of the lockplate and wide enough to allow passage of the lockplate screw.

The large branch of the sear spring must be well seated on the lockplate, and the moving one must have an almost imperceptible play; if it is too short the lock is hard to relax, and if it is too close to the eye of the sear, the same difficulty arises.

Since the springs are more fragile than the other parts of the lock, it is essential that they be of good strength.

The Screws (*vis*)

The stems must be well cylindrical and just right in the holes to receive them; their threads must be sharp and deep enough, the heads well-formed over and under, the upper edges a little rounded and well slotted in the middle; the slot should go only up to half the height of the head, which can separate when removing the screw if it is too deep.

To these might be added some observations by Dale on the inspection process:[46]

screws; if they have all their threads correct, their stem upright and well rounded; if the slot of the head is not too deep, which could make it split in two by the effort of the screwdriver.

wood screws: if their threads are full and well-formed without being sharp; if the pitch was not equal, the threads in the wood where these screws enter, would be cut every time the screws were inserted and soon there would be no threads at all.

trigger: if it has the shapes and dimensions that it must have, which is judged by eye.

Cotty concludes his details of the inspection process with the comment that: *There would still be many things to say about the mechanism of locks; but they depend more particularly on its geometrical outline, and we cannot speak of them without exceeding the scope of this publication.*

Annealing/Tempering

This process is described by Dale[47] as taking place only after the hardened pieces had been inspected and the lock parts, trigger, large lock and buttplate screws, and trigger guard screws, approved. Most of the hardened parts could not be used as they stood because they were too brittle, especially those of smaller size. This brittleness was removed by annealing, or tempering, which gave the iron some of the malleability lost in hardening. However, the lockplates were not annealed, the case-hardened layer being relatively insignificant with respect to their thickness; it was considered better to preserve on their surface all the hardness acquired in quenching.

The crest of the cock and the foot of the frizzen were annealed by holding them in melted lead, the cock's crest 5 to 6 seconds and the frizzen foot 13 to 14 seconds.

Screws, large and small, except the cock screws, were annealed by placing them, after rubbing them with oil, on an iron sheet and exposing it to the heat of a bright fire, and were removed when the oil was entirely consumed or evaporated, indicated by it no longer producing smoke. The screws of 40 guns were annealed at the same time.

The other small pieces were annealed after they had been cleaned to a bright finish by exposing them likewise, on a sheet of metal but without being oiled, to the action of a fire and leaving them there until heat caused the cock screw and the jaws to acquire a golden colour, and the other pieces blue.

The order in which each type of part took on the annealing colour was:

> the trigger, after 4 minutes, then the bridles, jaws, hammer screws, sears and finally the tumblers and top jaws which, being more massive than the other parts, were the last to succumb to tempering. The operation lasted 14 minutes.

The annealer straightened those lockplates which had been distorted during quenching (which he judged by looking at them in the direction of their length) and using a piece of walnut wood, slightly wider and longer than the lockplate, as packing, compressed them in a vice to straighten them.

To make sure that the annealing was neither too much nor too little, the annealer placed in each package one or two sample pieces of iron. After annealing, these samples were then broken, and the thickness of the case-hardening layer shown in the fracture cross-section indicated the degree of annealing.

The inspector in charge of the annealing tested all the large parts (lockplates, cocks, frizzens) one after the other, by trying to get a good file to 'bite'. He used a three-sided, so-called three-square, file. If it 'bit' in any place, it was a sign that that part had not been sufficiently hardened and it was necessary to case-harden again. In the same way, he tested as many other parts as possible.

It was felt better for the case-hardening to be a little weak than to be too strong, since, if the parts were too hard it would create fragility, reducing their serviceable quality. Also, better quality iron hardened more easily and more uniformly.

Blanc had gone a long way towards perfecting both the weapon and the approach to its manufacture, but being of the *Ancien Régime*, many of the mechanised methods were ignored in the republican zeal, and reliance was placed on hand-craftsmen once again. Even then, changes in the hierarchy did little to produce cohesiveness in the arms manufacturing establishment, despite, or perhaps because of, the creation of the *Atelier de Perfectionnement* in the early post-revolution period.

It is worth noting that in the early 19th century, John Bodmer, a Swiss civil engineer, also became involved in lock making. Little is known of his work beyond an intriguing sentence in his obituary[48] which stated he *used special machines for shaping the lock components with perfect uniformity*, and a note by Alder[49] that he had made interchangeable locks for the French Empire government within the 1800-1815 period.

Manufacture of the lock at Tula

Before examining lock-making practices recorded at Tula, it is instructive to compare the Tula Model 1808 musket lock with the French Model 1777 lock, which are, as can be seen, Figs. 5.33 and 5.34, to all intents and purposes, identical.

That two arms designers could independently create a lock in which there are so many elements of close correspondence is beyond coincidence, so it has to be concluded that the Russian M.1808 lock was a direct copy of the French M.1777.

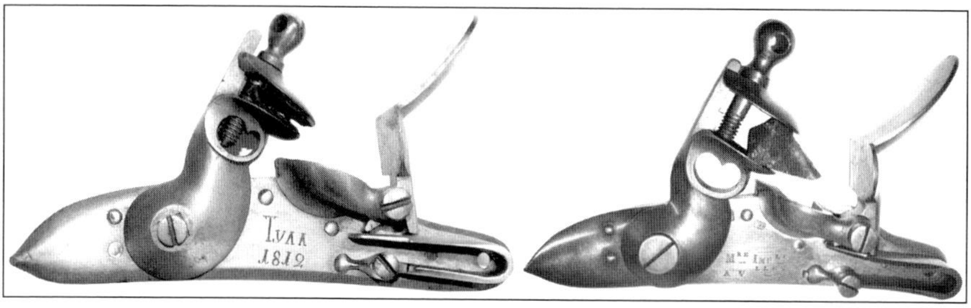

Fig. 5.33. Right: a French M. 1777 lock and Left: the Russian M. 1808 lock. (XII.2231 and XII.2351 respectively, © Royal Armouries)

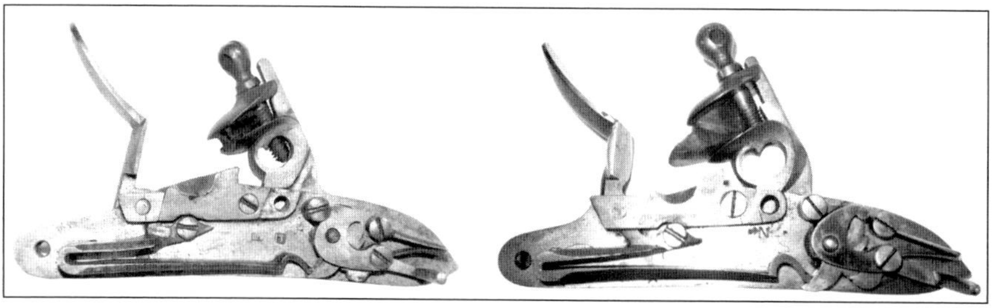

Fig. 5.34. Internal components of the locks in Fig. 5.33.

Fortunately, the making of the lock at Tula is explained in great detail by Gamel, and its similarity to both Blanc's original lock of 1777 and its later incarnations, gives some additional understanding of how that lock was produced and enables comparisons of the techniques and technologies employed.

Prior to 1819 at Tula the parts of the gun lock were forged by hand in the home workshops of the Tula craftsmen. These *had only a very little resemblance with the required parts with respect to their form* and were given *their necessary sizes by the fitters by means of tedious filing.* Around 1806, attempts were made to use a large screw press for die-forging, but it was *incomparable* to Jones's drop-forging process using dies, introduced in 1819, Fig. 5.35, the products from which *needed only a very small effort to be converted into perfectly finished interchangeable parts* [author's underlining].[50]

That the lock shown may be interchangeable is again demonstrated by the lack of assembly marks, placing it on a par with the Roanne and Liége locks.

It is difficult to imagine that the Jones', father and son, only developed these ideas after moving to Tula in 1817. It seems more likely that they took prior knowledge with them, gained in the gun trades in Birmingham – after all, John Jones the younger described himself as a Toolmaker and Diesinker in a patent

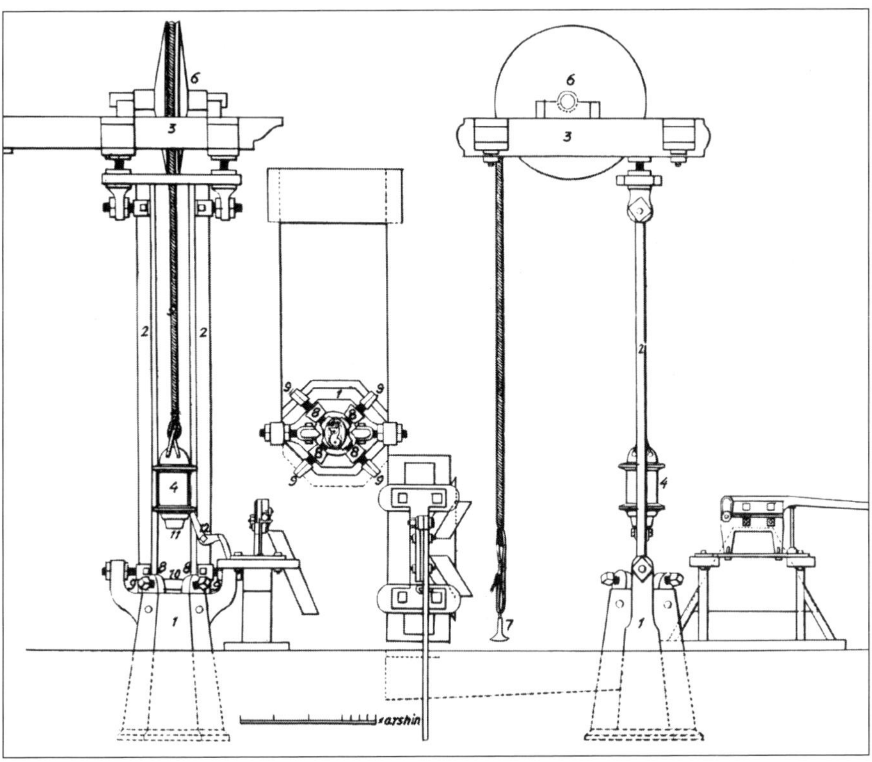

Fig. 5.35. Jones' drop-stamp at Tula used for die-forging lock components. (Gamel, 1826, Fig. XXVI)

Fig. 5.36. A drop forge in use at Tula - one operator placing the end of a bar into the lower die and two other operators, having hauled up the weight, are ready to let it fall. (Gamel, 1826, Fig. XXVII)

Fig. 5.37. Detail from Fig. 4, showing how the bottom die is positioned and fixed to the anvil by means of the large screws, 9.

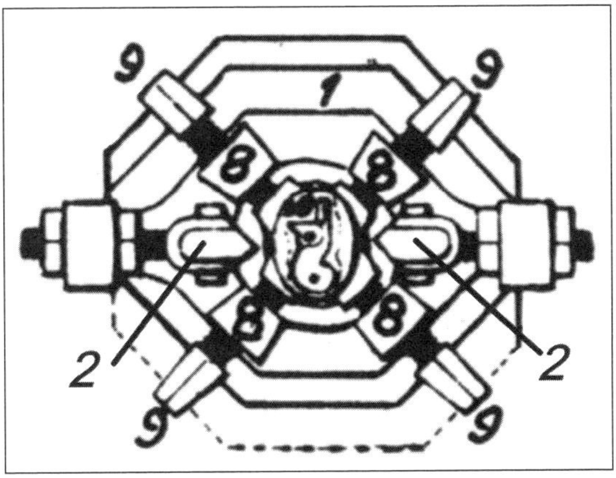

issued in 1806 - and is at least suggestive of die forging being in use at that time. Gamel claims[51] that whilst Jones had succeeded in forging flintlock cocks, he did not believe it was possible to die-forge the other parts. That he achieved this at Tula is indicated by the large range of dies and tooling used in the production of locks. However, a British patent issued to Round and Whitford in 1838 for die-forging gun components suggests that the Jones' never patented their methods in Britain.

The drop-stamp contains what have become traditional features. As shown in Fig. 5.35 a cast iron anvil, *1*, is set upon a good foundation in the floor. Extending upwards from it are two iron bars, *2*, *2*, with their inner edges shaped to form 'V's which act as guides for the falling weight, *4*. The weight, *4*, being two to four poods [33-66kg; see Appendix 1] is attached to a rope, *3*, passing over a pulley, *6*, at the top of the guides. Fig. 5.36 shows a drop stamp in use with workers hauling the weight up by hand to the top of the guides.

As in modern practice, the bottom die is fixed to the anvil by means of four large screws, *9*, Fig. 5.37, allowing it to be positioned accurately so that it mates perfectly with the top die attached to the falling weight. This feature is also the same as that described as being used in France.

The Lockplate

The two dies used in stamping the lockplate are shown in Fig. 5.38.

The piece of iron from which the lockplate was stamped was first heated and roughly shaped by hand forging, 'pointing' the rear end and reducing the width for a suitable distance to approximate to its narrower forward end. It was then reheated and laid on the bottom die which had been lubricated with hempseed oil. The hammer to which the top die was fixed was raised to the top of the frame by three men and allowed to drop. Having struck the iron in the die, it was raised and dropped again two more times and, in the intervals, the forging master relocated the iron forging in its correct position in the die for re-striking. Because of the displacement of the hot iron in this process, some was extruded between the dies to form 'fins' and these were trimmed off using another set of dies, as opposed to chisels and shears in French practice, in a 'fly press', Fig. 5.39.

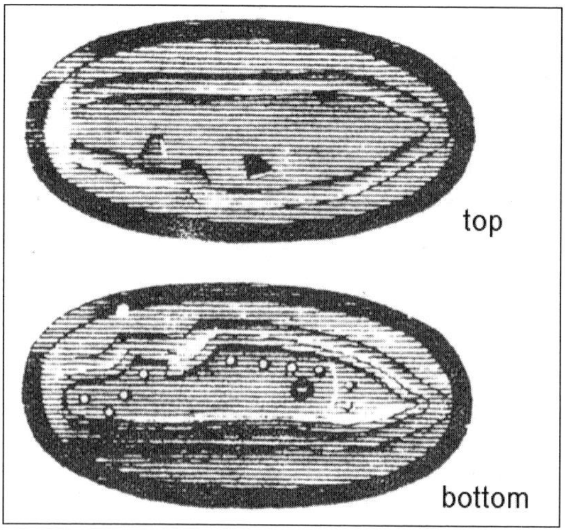

Fig. 5.38. The pair of dies used in forging the lockplate. (Gamel, 1826, Fig. XXX)

Fig. 5.39. Fly-press used for trimming lockplates. (Gamel, 1826, Fig. XXIX)

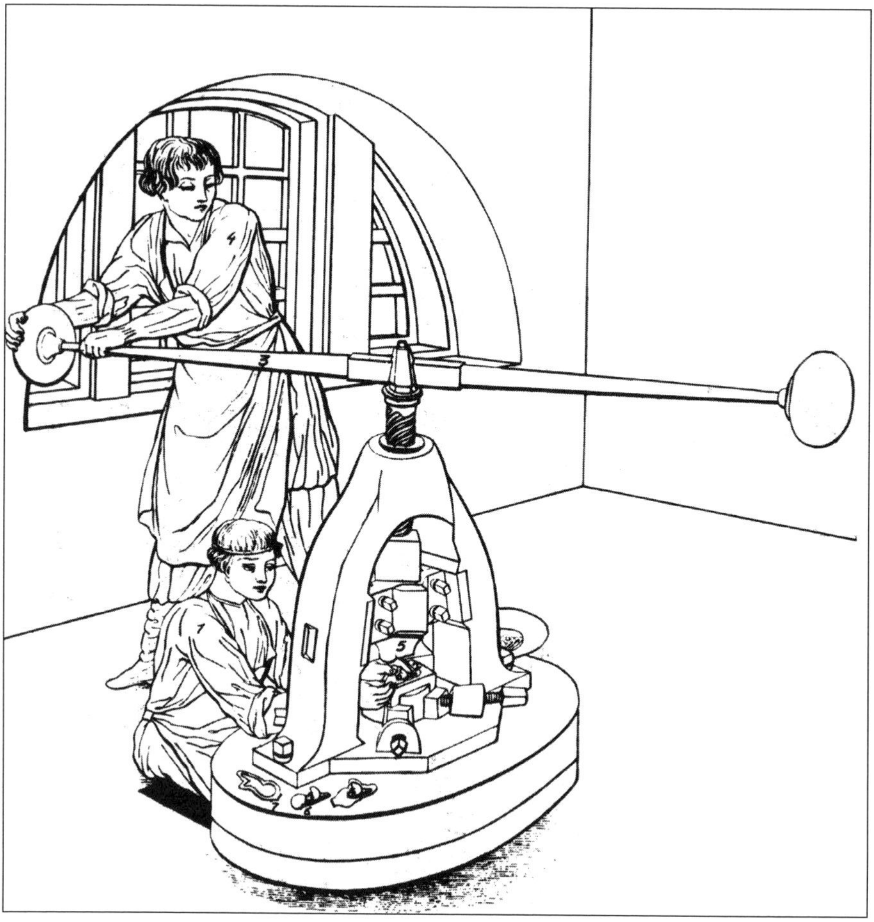

It will be seen that with the trimming dies shown in Fig. 5.40, the top die, (*4* and *6*), has a flat working surface but the bottom die, (*3* and *5*) is slightly 'dished' in part, allowing the edges of the dies to perform a shearing action on the 'fins'. This caused the lockplate to bend slightly, as it was forced through into the cavity in the bottom die. The lockplate was therefore heated again to red heat and returned to the forging dies for straightening with a single blow from the hammer.

Gamel notes that in the initial die-forging operation, since the iron was at a very high temperature it would have contracted slightly by the time it was trimmed in the cold state. In the straightening process therefore, which was carried out at a lower temperature than in the original forging, it was caused to actually fill the die and was thus brought to its correct size for the next stage. After removing the *fire scale*, left on the lockplates from heating and oxidation, by 'tumbling' them together with coarse sand and ashes, the hole for the large arbour of the tumbler was punched.[52]

Its position had been marked when forging and punching was carried out in a small screw or 'fly' press similar to that in Fig. 5.39. The punch tools for this process are shown in Fig. 5.41, items *7* and *9* being elevations and *8* and *10* plan views of the punch and die respectively. Although not mentioned, the die, *10*, appears to have a recess formed just below the hole, which may have been to accommodate the rear bolster [eminence] on the interior of the lockplate. After this initial forming of the hole by punching, the lockplate edges were, as in French practice, filed to the correct shape and size using a hardened filing jig. It will be recalled that at this stage in the production of the French 1777 lockplate it did not have any holes. The preliminary tumbler hole in the Tula lockplate could, therefore, have been used as a location and/or securing feature in fitting it to the filing jig, although this is not included in the description. However, such use might be precluded since the position of this hole was obviously not considered accurate enough.

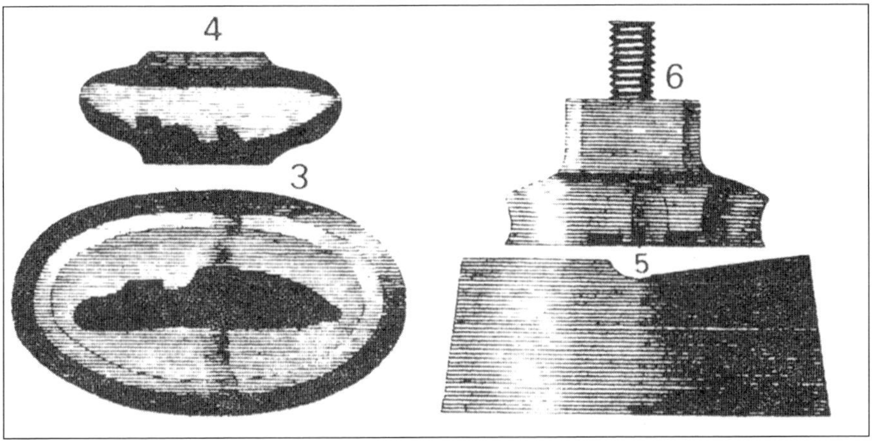

Fig. 5.40. Lockplate trimming dies shown in plan and elevation. (Gamel, 1826, Fig. XXX)

Filing the lockplate to size was an important preliminary to the next stage which was the drilling of the hole for the tumbler to give it its finished size and correct position. The lockplate was fitted into a special 'drilling fixture', Fig. 5.42, consisting of a plate, *2*, having a recess accurately matching the profile of the lockplate and into which the lockplate, *1*, was laid. It was then covered with another *reference* lockplate, *3*, in which a hole of the required size had been drilled to guide the drill in the drilling machine devised by John Jones in 1823,[53] Fig. 5.43.

The drilling machine, with which a man could drill 240 lockplates per day, consisted of a 'hand-brace' fitted into an iron frame and a table on which to rest the

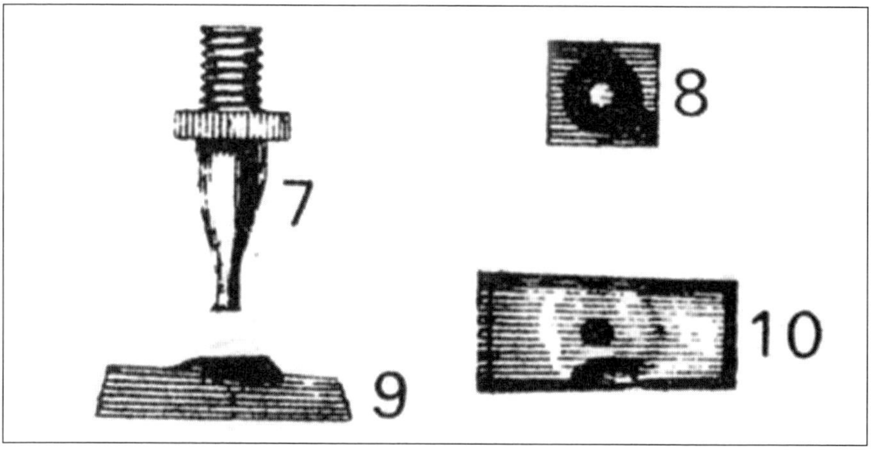

Fig. 5.41. Tools for punching the hole for the tumbler shaft in the lockplate. (Gamel, 1826, Fig. XXX)

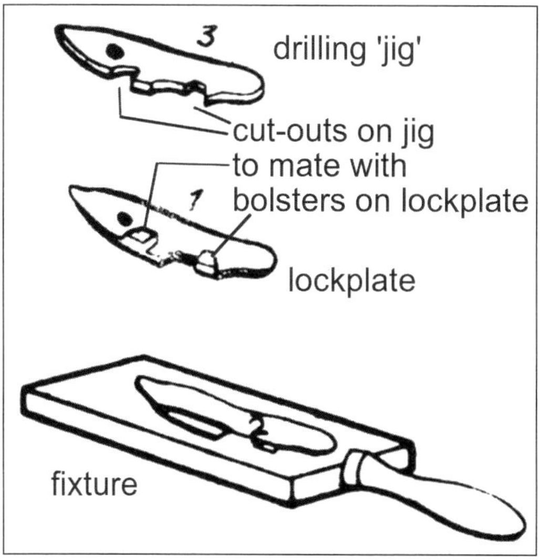

Fig. 5.42. Drilling fixture, 2, and reference plate 3, which acted as the 'jig' for drilling the tumbler hole in the lockplate, 1. (Gamel, 1826, rearranged Fig. XXXI)

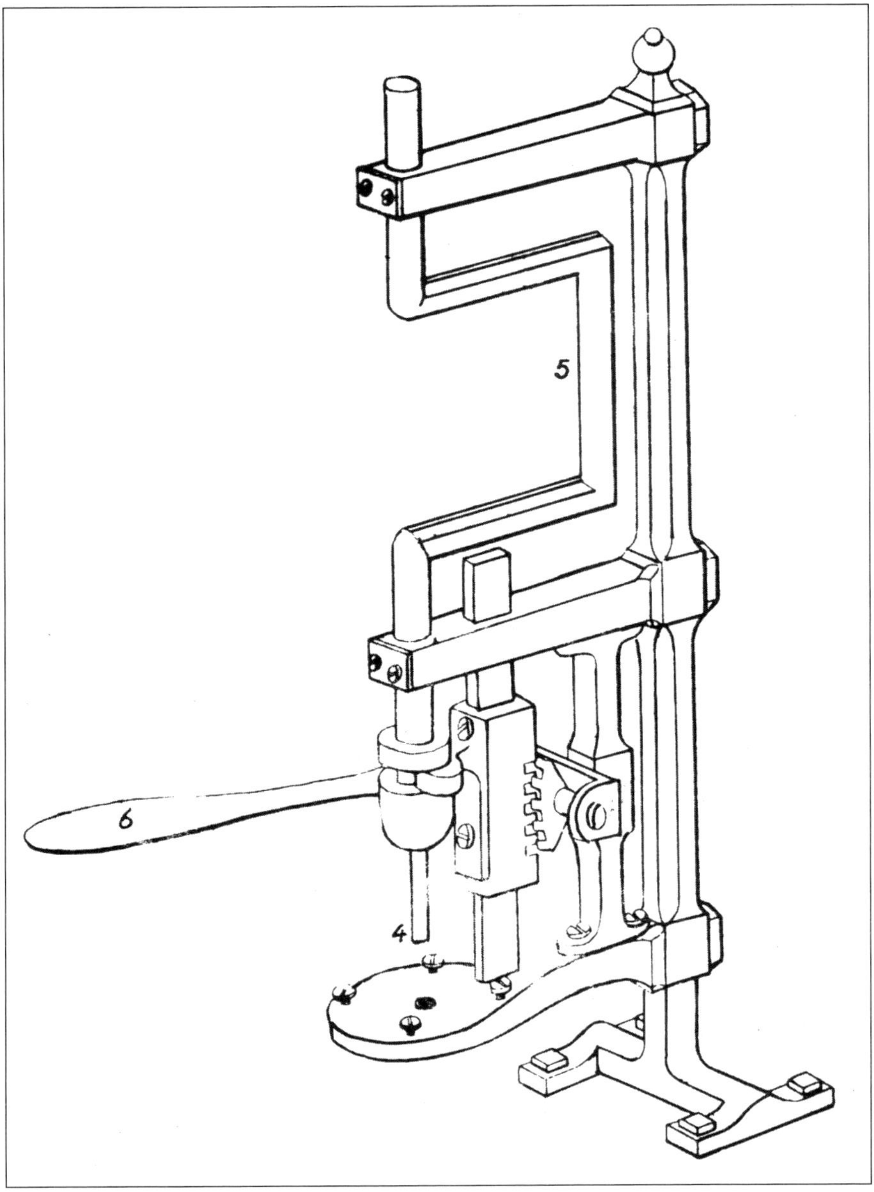

Fig. 5.43. Jones' tumbler-hole drilling machine. (Gamel, 1826, Fig. XXXI)

fixture and jig during the drilling operation; here again, this bears a resemblance to French practice. Bearing in mind that the hole had already been punched, presumably undersize, for this drilling operation to provide the correct result the drill needed to have a cylindrical body closely fitting the reference plate to prevent it 'wandering' off-centre and to cut only on its end face. This appears to be the

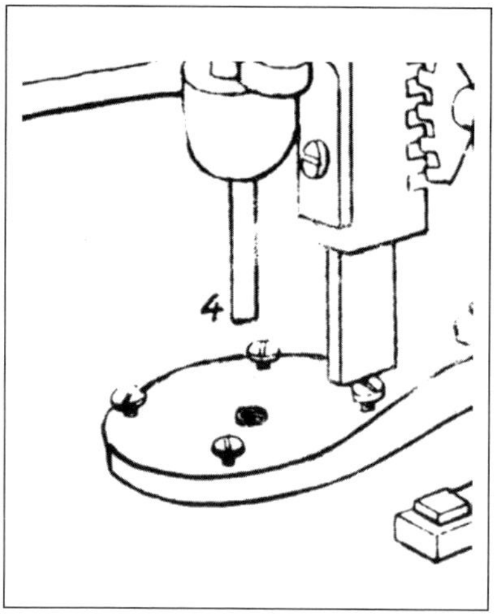

Fig. 5.44. Detail showing the drill, 4.

case on closely examining the machine, Fig. 5.44.

It will also be noted in Fig. 5.42 that the lockplate is fitted into the cavity with its flat outer face downwards to provide a flat seating. However, if the 'reference' lockplate, or 'jig' was simply a flat plate there would have been a risk of it canting over due to the presence of the bolsters – the 'trapezoidal eminences' of the French - on the inner side of the lockplate to be drilled. This would lead to an inaccurately drilled hole and it would seem sensible, therefore, to accommodate these projections in cut-outs on the reference plate, but there is no mention of such an arrangement.

When it came to drilling the remaining holes:

> *the lockplate is put in a reference die and covered by a reference plate in which all the holes required for the lockplate are made. A steel punch is used to mark the locations (where the holes are to be drilled) through these holes into the lockplate.*[54]

This would appear to be closely similar to the tooling shown in Fig. 5.42, but it begs the question of why this reference lockplate was not used as the drilling jig in this instance also? This question is not addressed by Gamel.

It also raises another question regarding the array of what are shown as circles in the bottom forging die, Fig. 5.45.

Could they be raised 'points' to mark the hole positions? It would certainly ensure that all the holes were marked in identical and correct positions, but again there are no comments by Gamel, so we are none the wiser. It is, however, worth noting that a similar practice existed in the Sheffield cutlery industry, and no doubt in similar industries also, in the manufacture of pen- and pocket-knives especially. To mark hole positions on the spring and the 'liner', which formed the casing of the knife, and in which holes had to coincide, a 'marking devil', Fig. 5.46, was used.

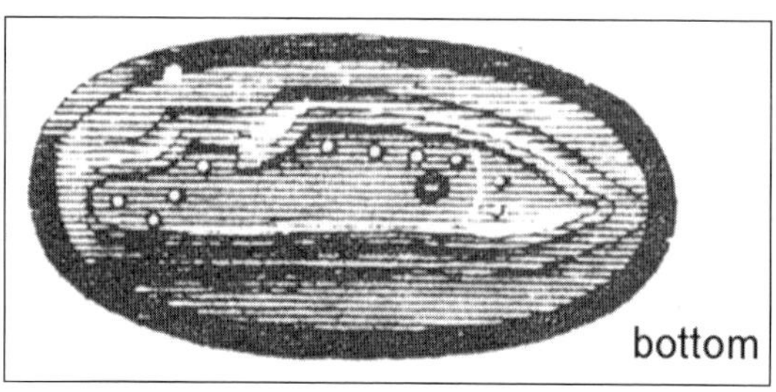

Fig. 5.45. The lockplate bottom die with unexplained small circular features.

The spring or the metal 'liner' was pressed up against the raised 'fence' and struck with a hammer, leaving centre marks in correct juxtaposition.

The lockplate drilling at Tula was carried out using a more sophisticated machine, Fig. 5.47, than the hand-brace used in France.

The machine had two powered spindles fixed perpendicularly to the worktables which were raised by levers, although the weights shown may have acted as a

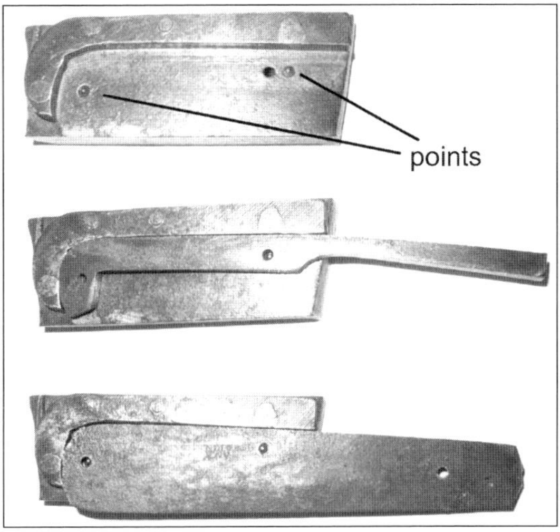

Right: Fig. 5.46. Sheffield pen- and pocket-knife cutler's 'marking devil' for giving repetitive accuracy of hole positions on springs (centre) and 'liners' (bottom). (Author's collection)

Below: Fig. 5.47. Lockplate drilling machine at Tula. (Gamel, 1826, Fig. XXXII)

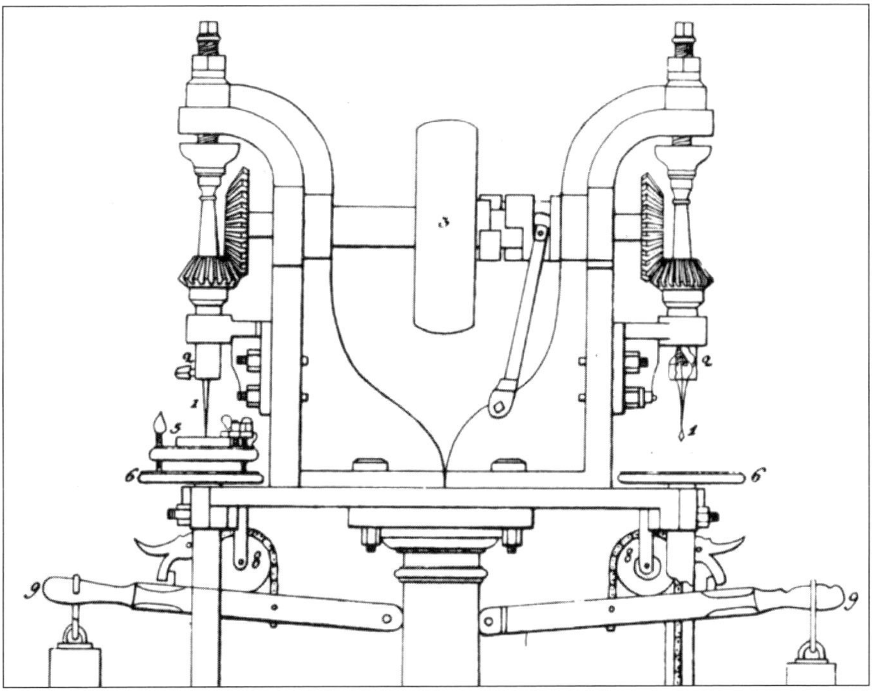

'power feed' to bring the lockplate into contact with the drill. The lockplate with its hole positions marked was fitted and clamped in a jig, Fig. 5.48.

This jig was used in conjunction with a brass 'register plate', Fig. 5.49.

This register plate was inset with two arrays of steel discs, *10*, each matching the array of holes in the lockplate, and each steel inset had a dimple in its centre, Fig. 5.49, and was fixed in a suitable position on the drilling machine table.

The jig, Fig. 5.48, had points formed on two diagonally opposing screws, A, Fig. 5.50, and it was used by placing these points successively in matching pairs of

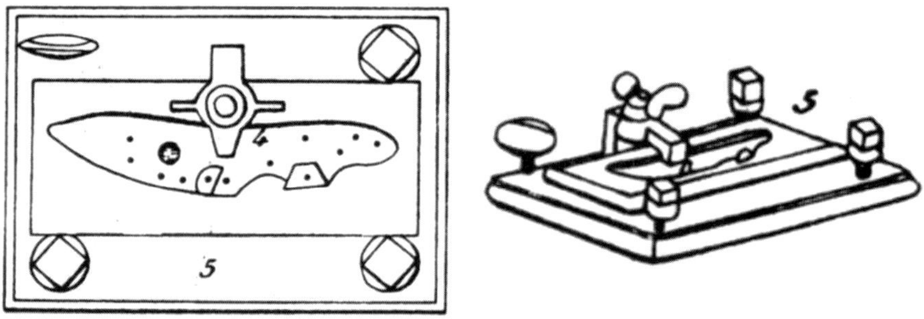

Fig. 5.48. The lockplate holding jig. (Gamel, 1826, Fig. XXXII)

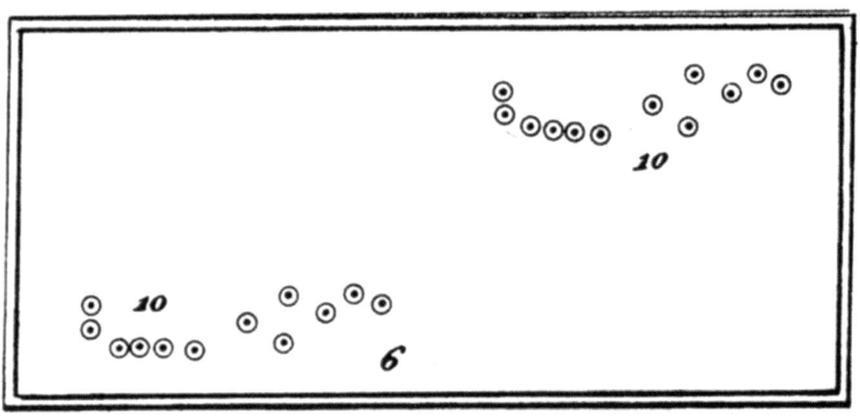

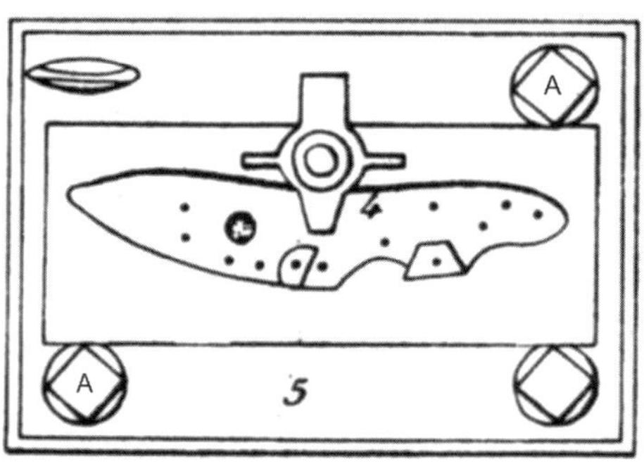

Above: Fig. 5.49. The 'register plate'. (Gamel, 1826, Fig. XXXII)

Left: Fig. 5.50. Showing the position of the pointed screws at A.

dimples on the register plate. It will be realised that, since the arrays of 'dimples' are diagonally opposed, only two diagonally opposed supporting screws on the jig could be fitted with 'points' since the others would have no dimples into which they could be located so may simply have acted as levelling screws.

Gamel illustrates the register plate and the jig side by side, Fig. 5.51, and it is assumed he intended to show their respective orientations in use – the position of the drill is shown by the 'X' added to Gamel's illustration – and each of the 'points' on the jig, on diagonally opposed screws, *A*, will engage with matching pairs of these dimples.

However, close inspection of the orientation given by Gamel in Fig. 5.51 shows it to be incorrect: placing the points of the jig in matching pairs of dimples on the register plate *does not* place the drill in the correct position on the lockplate to coincide with the hole centres! For that to happen, the jig as shown by Gamel has to be rotated through 180 degrees, Fig. 5.52:

> A simple demonstration shows this contention to be correct. Copies of Gamel's illustrations, adjusted as far as possible for scale, were transferred to pieces of card and holes punched where they were shown. These supplied models of the 'jig' and 'register plate'. A large black dot placed on the 'register plate', Fig. 5.53, shows where the drill would be centred.

The 'points' of the screws *A* on the simulated jig took the form of pegs attached on the reverse side, Fig. 5.54.

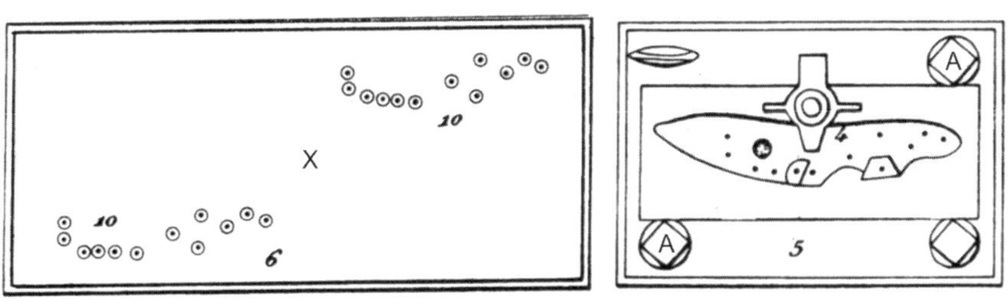

Fig. 5.51. Gamel's illustration of the register plate, 6, with the lockplate jig, 5.

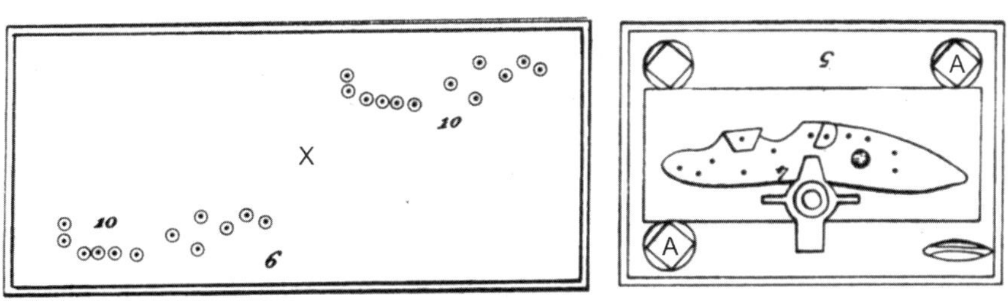

Fig. 5.52. Showing the correct orientation of the register plate and lockplate jig.

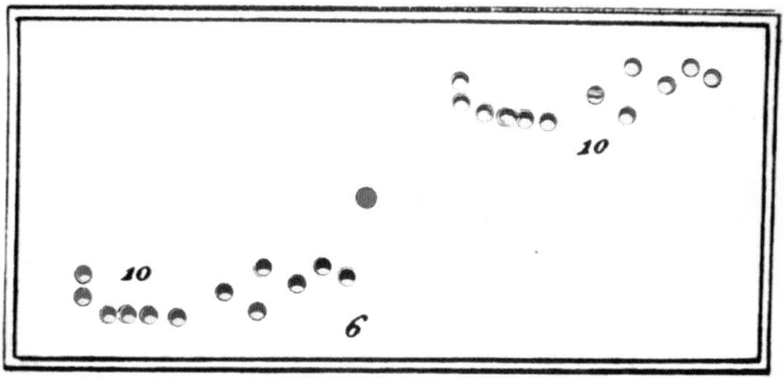

Fig. 5.53. Simulated 'register plate' with the drill position in black (© P. Smithurst)

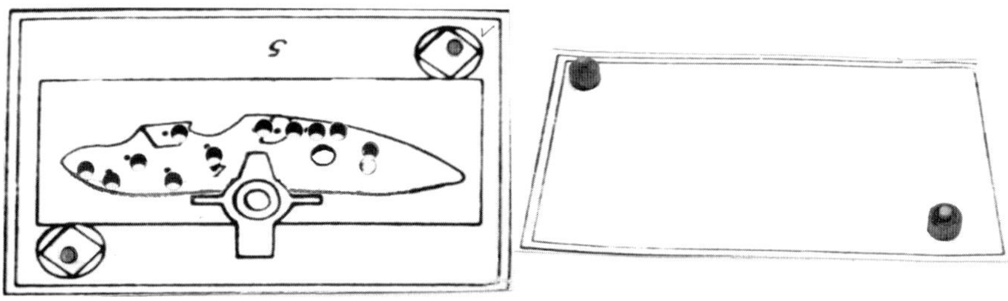

Fig. 5.54. Simulated 'jig' – top, left, and underside, right. (© P. Smithurst)

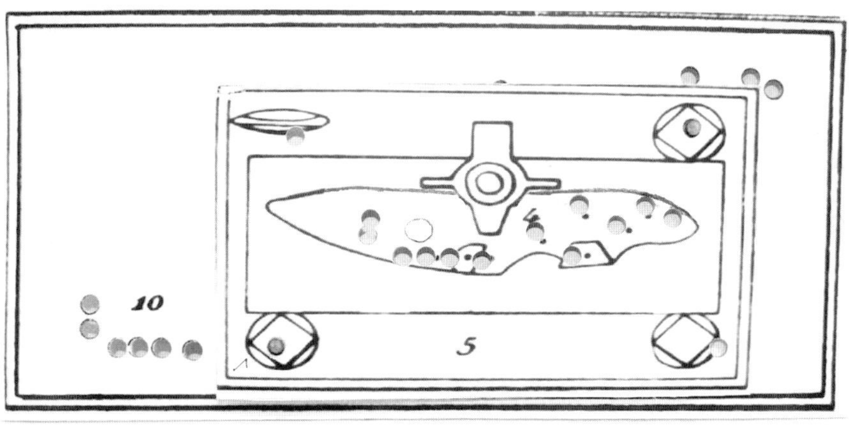

Fig. 5.55. In Gamel's orientation the drill and holes do not coincide. (© P. Smithurst)

If the pegs on the 'jig' are entered into corresponding holes on the 'register plate' in the orientation implied by Gamel in Fig. 5.51, none of the holes coincide with the 'black' drill position, Fig. 5.55.

However, by rotating the 'jig' through 180° it is found the hole and 'drill' coincide, evidenced by the black dot being visible through a lockplate hole, and it works with all other positions of the 'jig' on the 'register plate', Fig. 5.56.

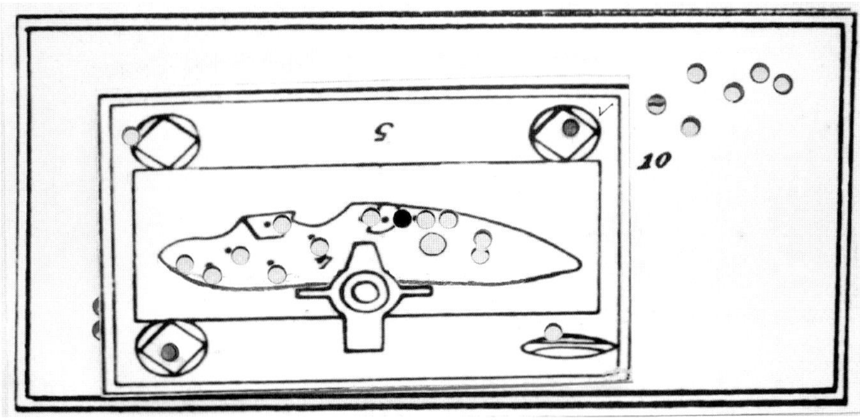

Fig. 5.56. Reorientation of the 'jig' on the 'register plate' gives coincidence of 'drill' and hole, evidenced by the visibility of the black dot close to the entre upper edge. (© P. Smithurst)

This was clearly a sophisticated technology, requiring high levels of accuracy - several identical arrays of holes had to be produced; the arrays on the 'register plate' had to be in exactly the correct mutual alignment; the inset steel discs with their 'dimples' had to be accurately fitted; the 'jig' holding the lockplate had to accurately position the lockplate in relation to the locating screws; these screws had to be accurately placed to bring the lockplate into correct alignment with the drill.

If, as seems possible, the lockplate forging die also marked the positions of the holes with centre-punch marks, then the points set in the die also had to be accurately positioned. Why it was used instead of the simpler method using a drilled and hardened lockplate as a drilling jig is not known, but it did, according to Gamel, achieve interchangeability.

After drilling, the lockplate was annealed ready for tapping – why it was not annealed for drilling is not stated – and the necessary holes tapped by hand in the conventional way, although there is no suggestion of any guide being used to ensure the taps were entered perpendicularly. This was followed by stamping the name and date of manufacture using a screw press and appropriate jigs and dies.

The Cock [Hammer]

The cock was die-forged using an asymmetric pair of dies, Fig. 5.57.

Fig. 5.37 actually shows the bottom die, *11*, in place but it will be noticed from Fig. 5.57 that the top die, *12*, only shapes the upper half of the cock. This was because only this portion extended beyond the vertical centre line, as may be seen from Fig. 5.58, and allowed simpler asymmetrical dies to be used.

The fins from initial forging were retained to assist locating the cock in the dies for punching the square hole to fit the tumbler arbour and the 'heart' shaped hole beneath the lower jaw. It will be seen from the forging dies, Fig. 5.57, that these holes were partially formed, or at least marked, at this stage so that punching these holes, using the dies in Fig. 5.59, did not have to remove the full thickness of metal.

Once this had been done, the fins were trimmed off using the dies in Fig. 5.60.

The hammer was then straightened in the same way as the lockplate by 're-striking' in the forging die.

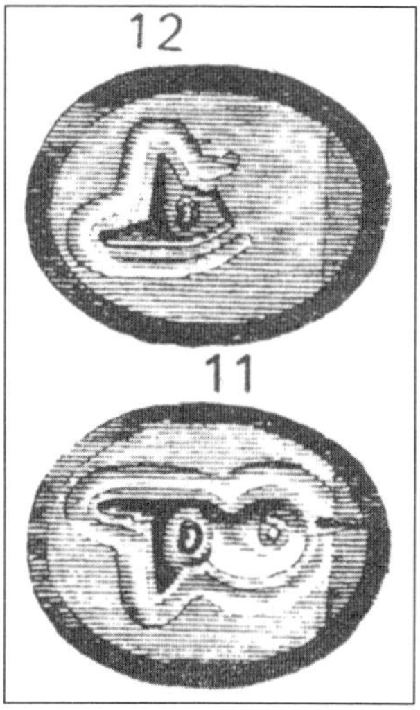

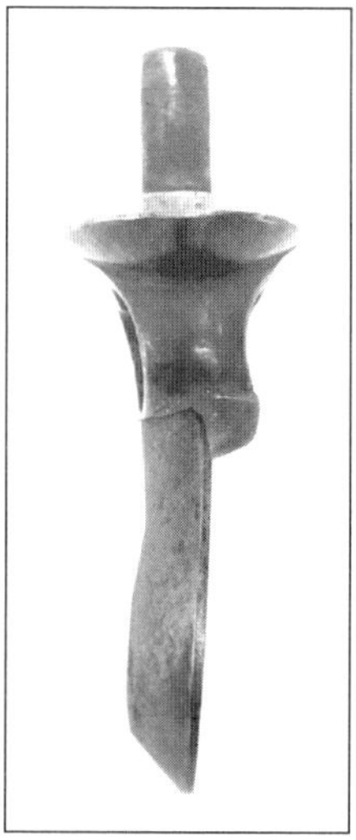

Far left: Fig. 5.57. Pair of dies used to forge the hammer. (Gamel, 1826, Fig. XXX)

Left: Fig. 5.58. Front profile of a slightly mis-shaped 1777/1808 cock. (Author's collection)

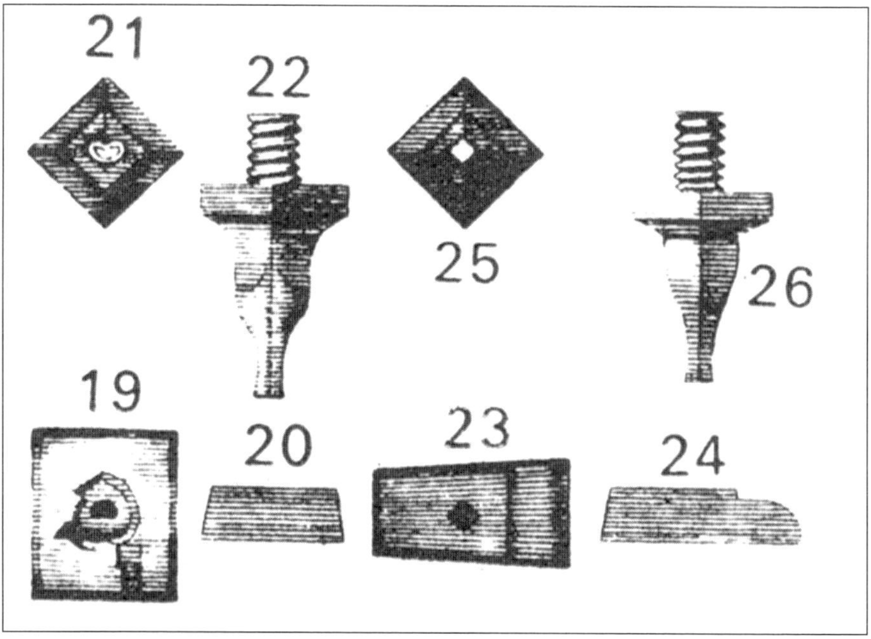

Fig. 5.59. Punches for creating the 'heart' and square hole in the hammer.

The next step was to form the threaded hole for the top jaw screw but there is some confusion in Gamel's account due either to the author or to mistranslation. It is stated that the hole for the top jaw screw was punched but that is contradicted by the statement that it was *fixed on a machine similar to that used in drilling the small holes in the* [lock] *plate*[55] and Gamel's illustration of the somewhat delicate and refined jig with the cock in place, Fig. 5.61, would suggest drilling rather than the much more brutal punching operation. The hole was then tapped.

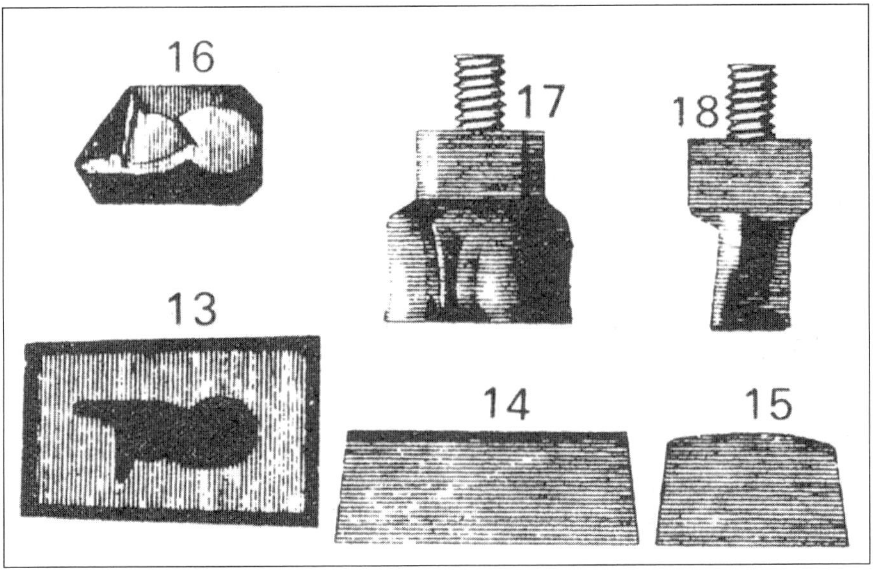

Fig. 5.60. Hammer trimming punch 16 and the die 13. (17, 18 and 14, 15 are side and end elevations of 16 and 13. Gamel, 1826, Fig. XXX)

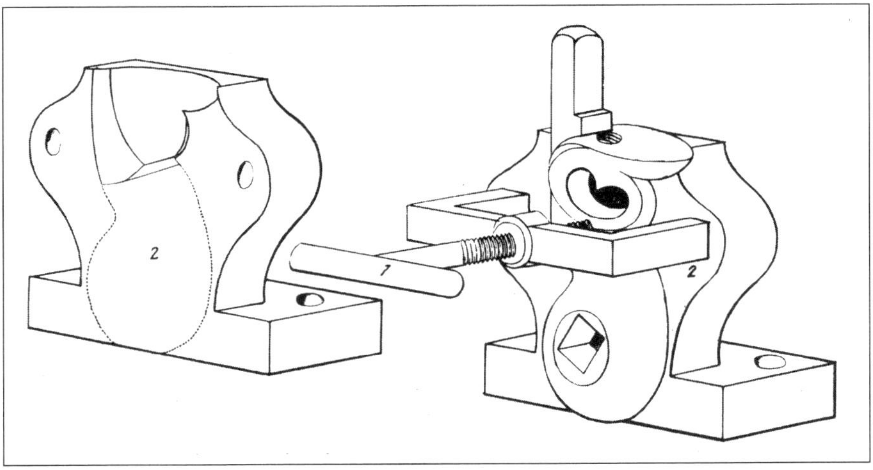

Fig. 5.61. Jig for drilling hole for top jaw screw. (Gamel, 1826, Fig. XXXIV)

The top jaw screw was forged by hand using dies.

The bottom die was fixed to the anvil while the top die is described[56] as *a moveable anvil which is fixed to the wooden lever with the other half of the screw cut in it is mounted* . This is somewhat garbled, probably in translation, and is an error - from both context and precedent *wooden lever* should read *wooden handle.* This view is supported by the evidence from the illustration of the dies in question, Fig. 5.63.

Item 65 in Fig. 5.63 is a side elevation of the top die and it will be seen to have a rounded 'collar' towards its top face where it is struck by the hammer. It would be logical to assume that this is the end of a handle – how otherwise could it be held in the correct position? Further supporting evidence comes from a simple exemplar of a set of dies matching the dies used at Tula.

In Fig. 5.64 is shown a 'bolster print', a tool used in the Sheffield cutlery forging trades for forming the 'bolster' – the swelling which separates the blade from the 'tang' upon which the handle is fitted on table and other knives.

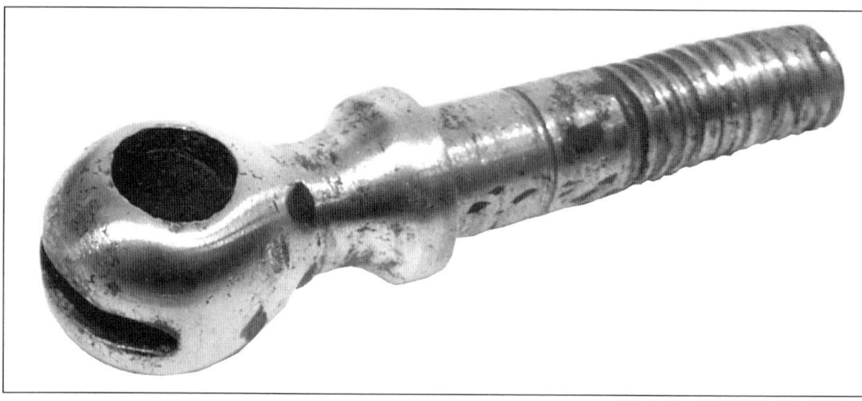

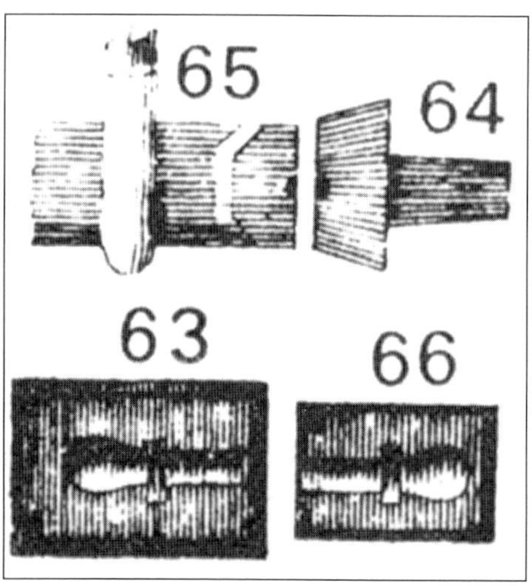

Above: Fig. 5.62. Typical top jaw screw with slot and a cross-hole for a 'tommy-bar'. (Author's collection)

Left: Fig. 5.63. The bottom dies, 63 and 64, and the top dies, 65 and 66 used at Tula for forging the top jaw screw. (Gamel, 1826, Fig. XXX)

Fig. 5.64. Set of 'bolster prints'.

The similarity is clear. As can be seen, these bolster prints took the form of a top die, usually held in a handle made from twisted hazel wands, and which neatly fitted into the guides forming part of the bottom die which would have been fixed in a groove on the face of the anvil.

In forging the top jaw screw at Tula, the master rotated the screw as it was being formed in the bottom die whilst the assistant repeatedly struck the top die. In this way an almost perfectly round screw head was created. The hole through the head of the screw was punched after heating and the slot was cut by a hand saw.

For turning the shank of the screw and threading it, two machines designed by John Jones in 1820[57] were used. Both allowed the use of interchangeable tooling to perform their functions on various components. The first machine, Fig. 5.65, was used to round the shank of the forged screw and machine the underside of the head square with the shank.

The machine consisted of a base plate, *1*, on which were mounted three cast iron supports, *2*, *4* and *8*. Two of these supports, *2* and *8*, had coaxial bearings fitted at the top and carried a hand brace, *6*. This, while being rotated by hand could be moved laterally by means of the lever, *15*, operating via sector gears, *12* and *13*, a forked arm, *10,* engaging with the groove in a boss, *7*, forming the front end of the brace. A hole in this boss allowed the fitting of a holding device, *16*, adapted to accept the head of the top jaw screw, *18*, which was secured by a pin, *19*, through its cross-hole.

In the left-hand column, *4*, the *hollow end mill, 5*, was fitted. An English 'milling cutter' of this type and size is shown in Fig. 5.66.

By turning the brace with the right hand and operating the lever, *15*, with the left hand, the rotating shank of the screw was fed into the hollow end mill and

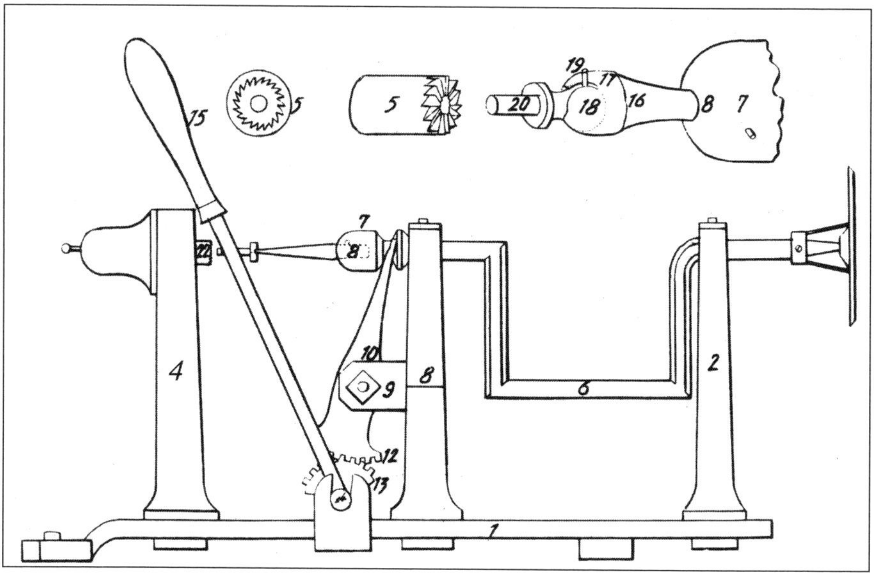

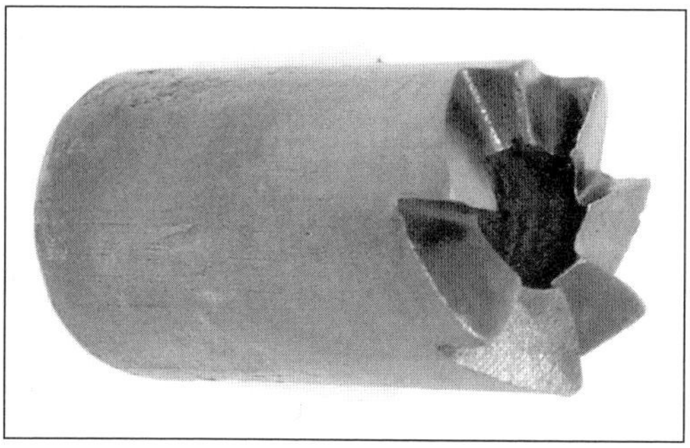

Above: Fig. 5.65. Jones' machine for turning the shank and facing the underside of the head of the top jaw screw. (Gamel, 1826, Fig. XXXV)

Left: Fig. 5.66. A hollow end mill, similar to those used at Tula and in France, from the English gun trade. (Author's collection)

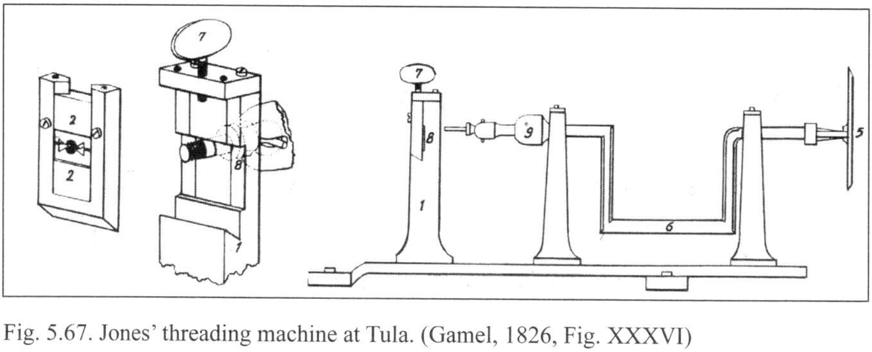

Fig. 5.67. Jones' threading machine at Tula. (Gamel, 1826, Fig. XXXVI)

turned to size by the sharp corners of the end mill at their junction with the hole. When the shank was fully inserted, continuing the rotation of the hand-brace faced the underside of the screw head square with the shank.

Threading was carried out on a similar machine, Fig. 5.67, which was used for all the lock screws.

The same device holding the screw by its head was fitted into the boss, *9*, of the hand-brace, *6*. The column, *1*, was adapted to take a 'guide plate' having a hole to match the shank of the screw. The screw, passing through this hole, entered the opening in a pair of dies, *2*, located in the 'V' guides of a housing which in turn was fitted to the column, *1*, and secured by two screws at its top. With the machine fixed to a bench, the hand-brace was operated in conventional fashion by being pressed against the breast of the operator whilst the handle was turned, causing the screw shank to enter the dies for its thread to be cut. The correct positioning of the dies to cut the necessary depth of the thread was accomplished by the thumb screw, *7*. These dies correspond closely to those used in some larger diestocks.

The Tumbler [Trunnion]

Making the tumbler, referred to as a *trunnion* in the translation,[58] was an elaborate process befitting its elaborate nature. It was initially hand-forged to the approximate shape, Fig. 5.68.

Gamel[59] reports that until 1826 tumblers were machined in what he refers to as an *English device*, Fig. 5.69, which was a 'tumbler mill'.

The mill was fitted with two opposed 'facing' cutters, the lower one of which can be seen in Fig. 5.69, and once assembled in the mill, the protruding tip of the larger shank of the tumbler projected through the bottom cutter, *7*, and mounting plate, *2*.

This tip was gripped in a vice and, with reference to Fig. 5.69, the mill was rotated by its handle, *4*, whilst the facing cutters were gradually moved closer together by the

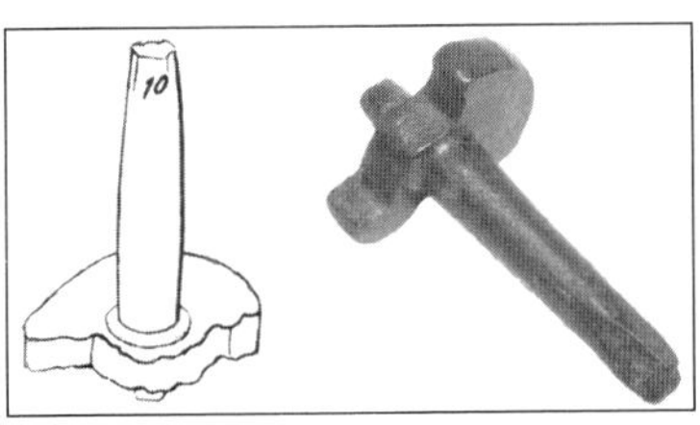

Above: Fig. 5.68. The Tula tumbler forging alongside an later English tumbler forging. (Gamel, 1826, Fig. XXXVII)

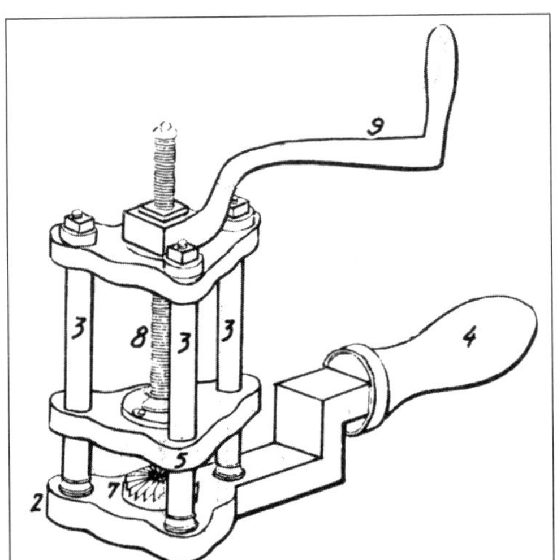

Right: Fig. 5.69. The 'English device' – a *tumbler mill* - used at Tula. (Gamel, 1826, Fig. XXXVII)

lever, *9*, which rotated a 'feed screw'. In this way, the shank, the small 'spigot', and, when the cutters had passed down the shanks, the faces of the tumbler were machined. The faces of the tumbler body were thus parallel and the shank and 'spigot' coaxial and square with the faces of the body – essential features in a tumbler.

Shortly before 1826,[60] a powered machine was invented to perform these operations, Fig. 5.70.

It utilised the same principles of two 'facing' cutters being brought together to machine the face of the tumbler body and to round the shank and pivot. The major difference was that the tumbler, not the cutters, was revolved through the drive belt and gearing arrangement. Also, the rotation of the tumbler could be reversed. The reason for this may be because the cutters had teeth facing in opposite directions so one will cut in clockwise rotation while the other simply rubs against the face of the tumbler. Considering the tumbler was made from bayonet steel,[61] such 'rubbing' was likely to cause noticeable wear of the cutter teeth.

Finally, the tumbler body needed accurate shaping. Tula was again indebted to Jones for a simple and effective way of correctly marking the shapes that needed filing. A press was fitted with a pair of matching dies, Fig. 5.71, each having a raised steel 'knife-edge' formed accurately to the tumbler profile and great care had to be taken to ensure they were perfectly aligned with each other.[62]

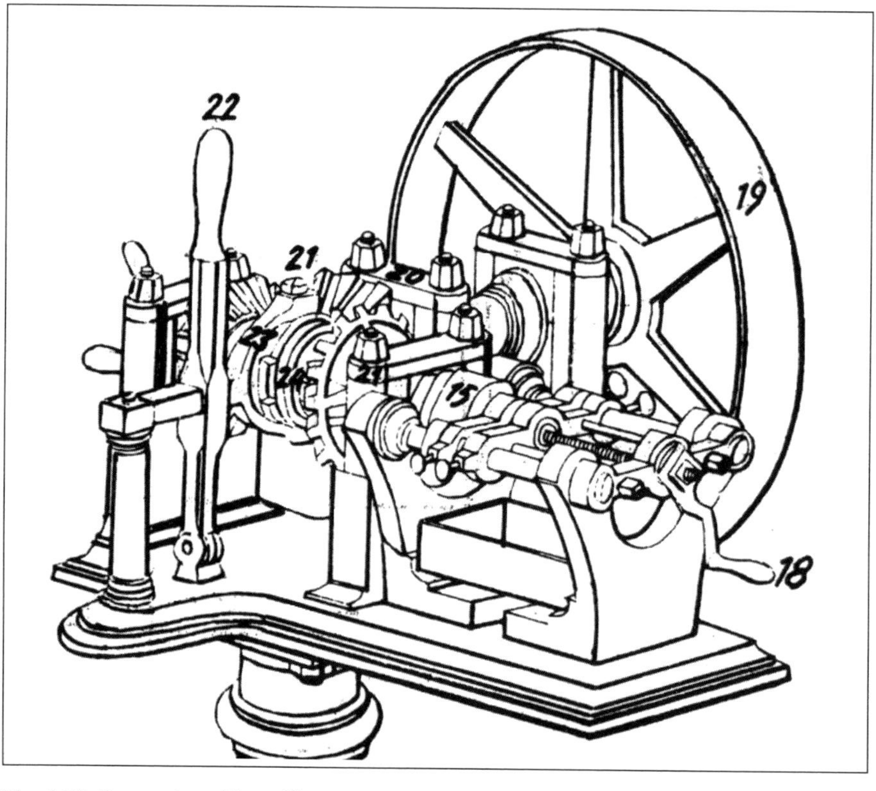

Fig. 5.70. Powered tumbler mill. (Gamel, 1826, Fig. XXVII)

Once the profile had been marked, the tumbler was then carefully filed to shape but, unlike Blanc's technique used in France, it was not then forced through a 'finishing die' to give it the correct size and form.

The next crucial operation on the tumbler was filing the square on the large shank onto which the cock fitted. Every musket lock had to have exactly the same relationship between the positions of the tumbler, the sear and the hammer/cock. Again, we encounter Jones' input into this procedure by devising a complex two-part filing jig, Fig. 5.72, in which the tumblers were grasped in identical conformation so that the guides for the files produced the square with repeatable accuracy.

The tumbler was fitted into a slot and grooves matching its body and shanks in one half of the jig which had an array of locating pins, *14*. The second half with an array of holes, *15*, matching the locating pins was also fitted with a 'cradle', *17*, having a cut-out, which was a female version of the tumbler profile, incorporating a 'tooth', *18*, shown in detail in Fig. 5.73.

When assembled together this tooth, which is actually a 'step', engaged with the 'firing' notch on the tumbler, ensuring it was correctly positioned and prevented

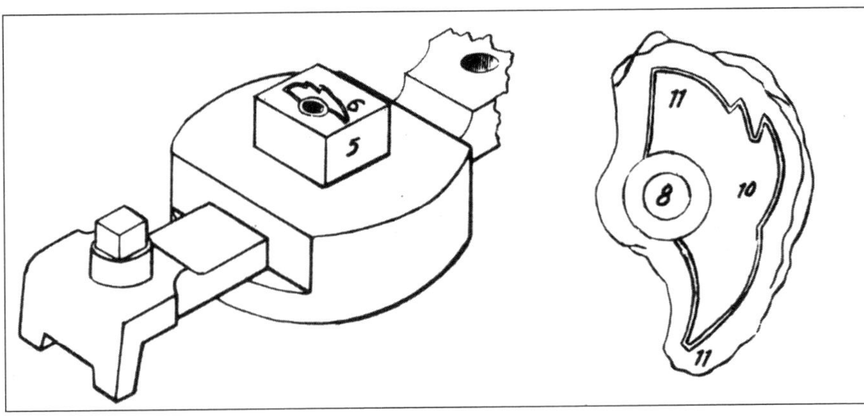

Fig. 5.71. The bottom die used to mark the correct profile on the tumbler and, right, the tumbler body with the profile incised. (Gamel, 1826, Fig. XXXVIII)

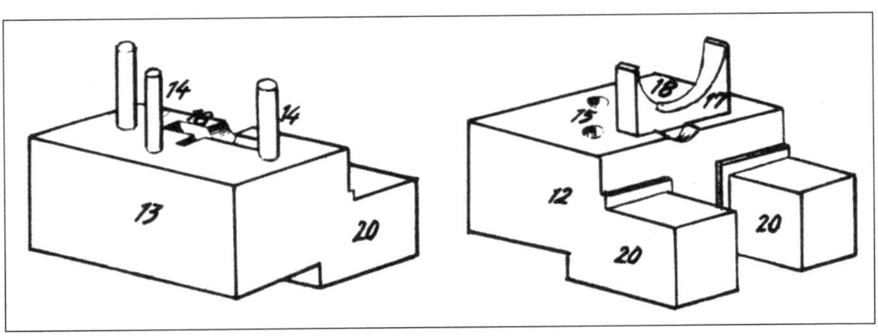

Fig. 5.72. The two parts of the jig for filing the square on the tumbler shank. (Gamel, 1826, Fig. XXXVIII)

any rotation. When the two halves were assembled, Fig. 5.74, the large shank of the tumbler projected into the cruciform trough formed by the 'legs', *20*, of each of the two halves of the jig and these 'legs' then acted as guides for the file. This may approximate to Alder's *calibre double* mentioned earlier.

There is a conundrum with every filing jig: if it is assumed the legs, *20*, would need to be hardened to avoid damage, how, then, was file wear, and associated inaccuracy, controlled? If the legs, *20*, were not hardened, then a gradually increasing divergence from standard would result! These would be guaranteed to negate the very purpose of the jig but are questions neither raised nor answered by Gamel.

Jones also devised a simple machine for drilling the hole in the end of the square for the cock-screw. Prior to this, Gamel observes that these holes usually ended up inclined, which would cause the cock-screw to only bear on a small section of its outer edge. The machine, Fig. 5.75, consisted of a frame with a simple 'headstock' and 'tailstock'.

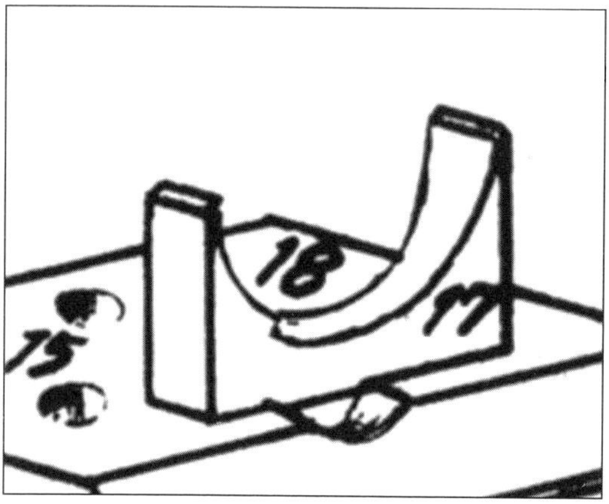

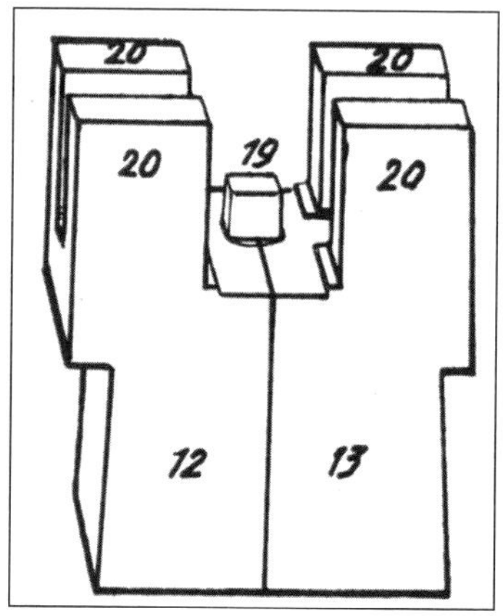

Above left: Fig. 5.73. The tooth, 18, in the 'cradle', 17.

Above right: Fig. 5.74. Assembled filing jig for the square on the tumbler's large shank. (Gamel, 1826, Fig. XXXVIII)

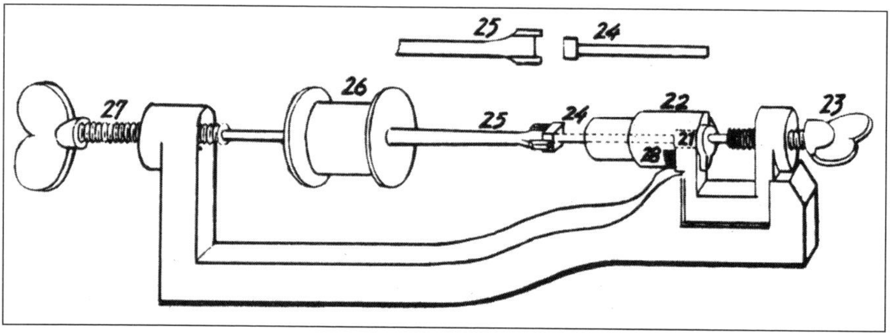

Fig. 5.75. Jones' machine for drilling the hole in the tumbler square. (Gamel, 1826, Fig. XXXVIII)

The tailstock carried a cylindrical sleeve, *22*, into the right-hand end of which the square on the tumbler fitted accurately and was held in place by a thumbscrew, *23*. The 'drill', *25*, was a typical 'bow drill' in construction but had a screwdriver blade with a projection at each end. This engaged with a slot in the square head of a drill, *24*, the projections preventing it from sliding out of the slot during use. The shank of the drill closely fitted the hole in the extended nose of the cylinder, *22*, which centred it on the tumbler square fixed at the opposite end. The left-hand end of the drill shaft had a point which fitted into the adjusting screw, *27*. Operating the bow with the right-hand caused the drill to rotate in alternate directions, and the left-hand was used to advance the drill into the tumbler square by means of the screw, *27*. How depth was controlled is not stated but it would have been a simple matter to fit a sleeve of pre-determined length on the shank of the drill, *24*, for instance. Once the hole had been drilled, it was tapped by hand.

The Bridle

The bridle deserves special mention because of its intimate association with tumbler, sear and lockplate; while location of all other holes on the lockplate was important, those for tumbler, bridle and sear were vital for the lock to function correctly, Fig. 5.76.

No illustration of the bridle is supplied by Gamel but being a copy of the French M.1777, one from that musket is shown in Fig. 5.37.

As previously noted, the important feature of the bridle was the hole on its face into which the small pivot of the tumbler fitted. This had to be coaxial with the large tumbler hole in the lockplate to maintain parallelism of the tumbler body with the inner faces of the bridle and lockplate. Any divergence would result in the tumbler 'binding' in the holes or rubbing against those faces and interfering with the smooth functioning of the lock. This meant that all the associated holes in the lockplate, including the one for the locating pin on the bridle, had to be coincident with those of the bridle. Its whole production process was another of Jones' contributions.

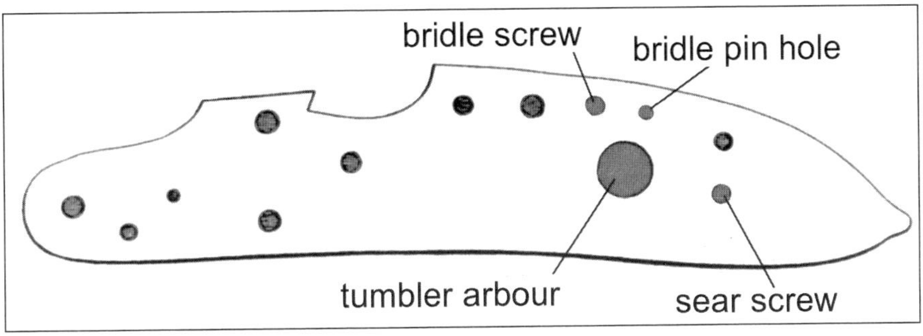

Fig. 5.76. The inner face of the M.1808 Tula lockplate showing the holes whose relative positions were critical. (© P. Smithurst)

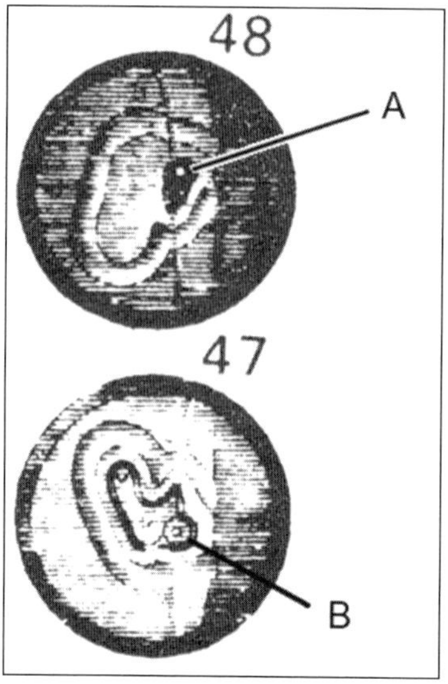

Fig. 5.77. Forging dies for the bridle which may incorporate markers, A and B, for the hole for fixing screw and the locating pin. (Gamel, 1826, Fig. XXX)

The bridle, like other components, was forged in dies, Fig. 5.77.

Two small features in these dies, *A* and *B*, have been highlighted. As with the lockplate die, they are suggestive of 'points' or other devices to mark the positions of the hole for the fixing screw in the 'foot' of the bridle and that for the locating pin. There is some uncertainty about this locating pin. Gamel refers to an *internal pin* as a means of locating the bridle, but this might be a misinterpretation and could be meant to be 'integral', which would accord with what is believed to have been common practice.

The hole for the bridle screw was drilled in a machine similar to that used for drilling the lockplate and for this, the bridle was placed in the jig, *5*, and clamped by the screw, *6*, shown in Fig. 5.78. For the bridle to sit firmly on the face of the jig, the jig would need a recess to receive the 'foot' of the bridle, and this is visible as the heart-shaped cavity. It is clear that, if as assumed, the bridle had a location pin – an essential feature – then it would have needed to have been formed before this operation could be carried out and it would need to fit in a hole at the bottom of the heart-shaped cavity for the foot of the bridle in the jig.

Drilling the hole for the bridle screw first was important when it came to drilling the remaining holes for sear screw and tumbler pivot. Their locations were equally critical. For this, the lockplate itself was used as a drilling fixture. Two drilling jigs/guides were used, Fig. 5.79.

As can be seen in Fig. 5.80, the bridle, *8*, was fitted to the lockplate using its locating pin and its fixing screw, *7*.

For the drilling jig, *9*, to function, the lock would have to be turned over and the bridle drilled from the outer face of the lockplate, but exactly how that was

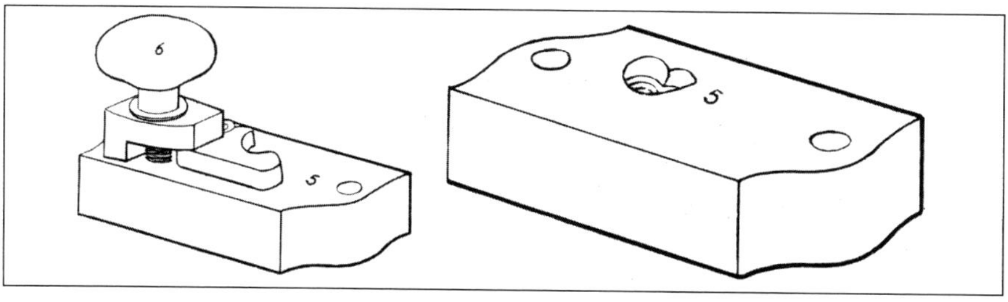

Fig. 5.78. Bridle-screw hole drilling jig. (Note that in the right-hand drawing the threaded hole for the clamping screw is missing or mis-placed) (from Gamel, 1826, Fig. XXXIV)

accomplished is not described. Also, there is no indication of how the jig and drill, *13*, for the sear screw hole in the bridle, is located other than it being secured by a screw from the outer side of the lockplate, presumably using the hole for the sear spring screw.

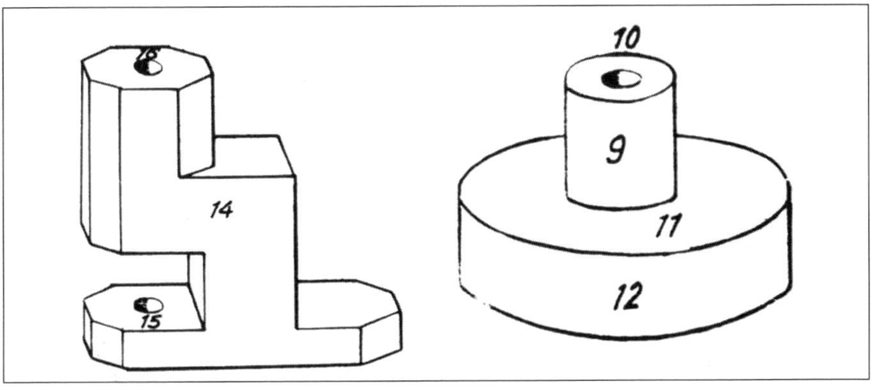

Fig. 5.79. Drilling jigs/guides used in conjunction with the lockplate for drilling holes in the bridle. (Gamel, 1826, Fig. XXXI)

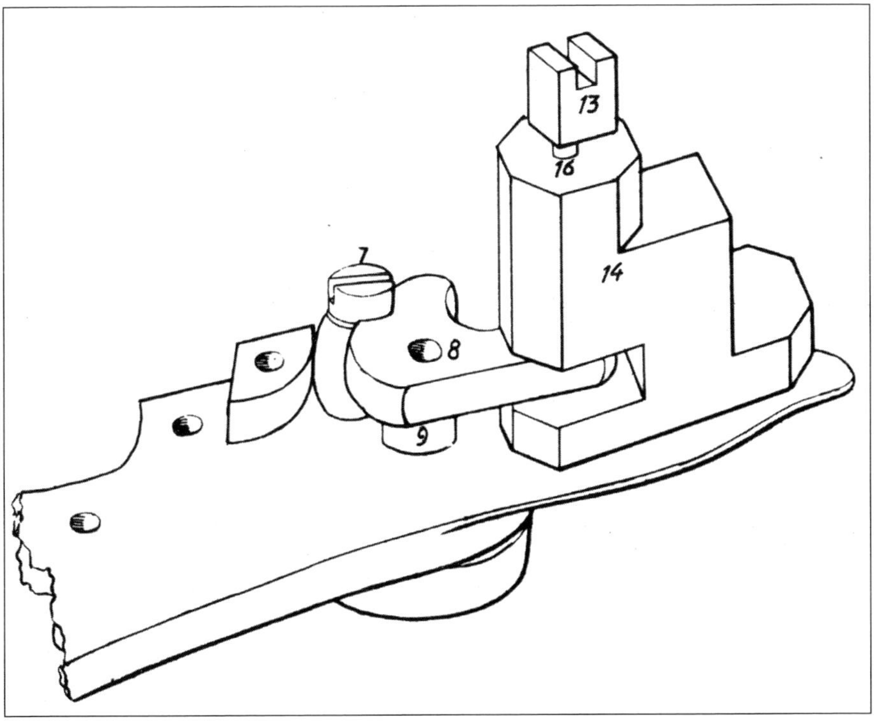

Fig. 5.80. The bridle fixed to the lockplate and the two drilling jigs in place. (Gamel, 1826, Fig. XXXI)

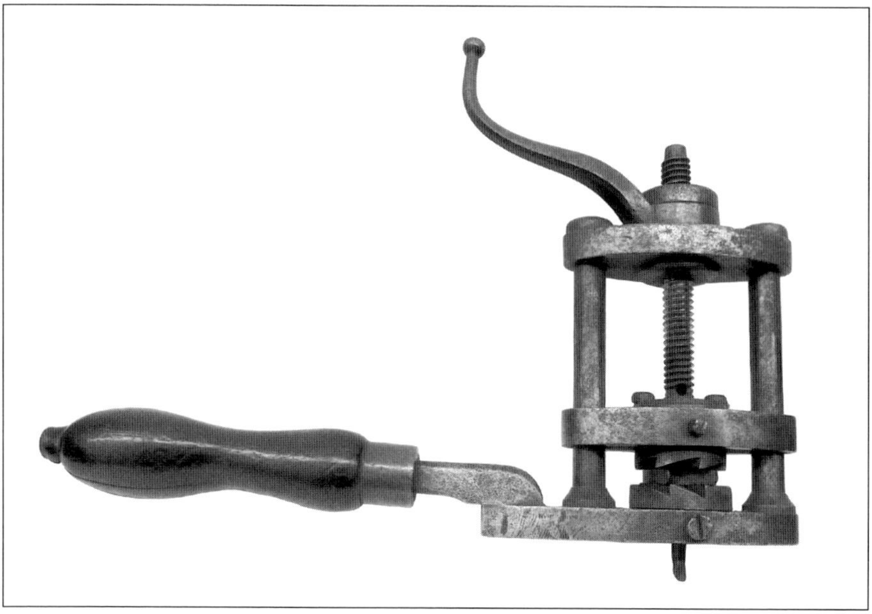

Fig. 5.81. Tumbler mill made by Whitehouse, a well-known tool maker to the gun trades. (Author's collection)

5.3 Lock making in Britain in the early 19th century

Sadly, within this period not even scant details of lock-making practices have been found. The Report of the Select Committee on Small Arms in 1854 notes the absence of any mechanisation but at least one exception was the tumbler mill, an example of which is shown in Fig. 5.81.

Beyond that, it can only be conjectured that practices were more on a par with those of France than those in Tula.

5.4 Lock making in America in the early 19th century

In 1794, with the prospect of a war with France looming, Congress authorised the establishment of National Armories for the manufacture of arms, and these were located at Springfield, Massachusetts and Harper's Ferry, Virginia. At the same time, when an amicable relationship with Britain had been re-established, and when, as Eli Whitney commented, gunmaking skills in the United States were insufficient to meet the needs of supplying an army,[63] complete arms and many components were purchased in Britain, including 6,000 locks for rifles and muskets from John Ketland.[64] This was a stop-gap measure – Springfield Armory was still getting underway and only produced 245 muskets in 1795 and none would be produced at Harper's Ferry until 1801. In 1798 the United States' Secretary of War,

James McHenry, established uniformity as a procurement goal for American arms makers[65] and in the same year, contracts were placed with twenty-seven contractors for 40,200 muskets of the French 'Charleville' model of 1763.[66]

One of these contractors was Eli Whitney, who had the largest contract for 10,000 muskets. He was not a gunmaker and set about designing a series of machines to perform individual sequential operations on the basis of 'one machine for one job'. Although he appears never to have used that exact phrase, it is a tenet implied in some of his and other's statements.[67,68] He has even been labelled as the pioneer of this approach[69] although the same system was used by Bramah and Maudslay around 1790[70,71] for the manufacture of components for Bramah's patent lock, and around 1805 by Brunel and Maudslay for the manufacture of ships blocks.[72] The same approach is even found in a screw-making machine[73] which became the basis of an operational factory in 1762.[74] It has also been suggested[75] that in his venture Whitney achieved interchangeability, a fact apparently attested to by a letter from Jefferson.[76] However, it has been suggested that at the time Whitney was working, sufficiently precise tools did not exist in the United States to enable interchangeability to be achieved,[77] a view reflected by Gordon[78] in an analysis of excavated material from Whitney's factory site. This view is borne out by the examination of a lock of the French 1777 style made by Whitney at this time in the Springfield Armory collections which bears the 'assembly marks' common to non-interchangeable locks.

Nevertheless, during the course of this contract, he saw an example of the French Model 1777 lock, very likely one of those procured by Jefferson during his role as U.S. representative in Paris. It is abundantly clear from a surviving letter written by Thomas Jefferson, the American representative in Paris, to John Jay, Secretary of Foreign Affairs in 1785,[79,80] that it was Blanc's work and his 1777 musket that provided the inspiration for the pursuit of interchangeability in America, although this letter has recently been incorrectly ascribed[81] to Benjamin Franklin. Depending on which account is read, in 1789 Jefferson sent either half a dozen 1777 muskets[82] or half a dozen of the 1777 locks[83] to America.

Impressed by the advantages of the Model 1777 lock over the Model 1763, Whitney wrote to Samuel Dexter, Secretary of the Treasury, suggesting the current musket be altered to adopt these advantages:

> *Improvements in the construction which in my opinion may be adopted with advantage. The modern French locks have a brass pan, which has the advantage of not being corroded by the nitre of the powder, of being placed with greater facility in a more favourable position to receive the Fire from the hammer* [frizzen], *and of more effectually securing the priming from the rain – which advantages I think will more than compensate for the expense of the alteration. I will contract to make Brass Pans for the second parcel of 1500 stands for six cents each in addition to the contract price, and Brass Pans with an alteration in the hammer* [frizzen] *etc. so as to embrace all the advantages, to the remaining seven thousand stands for an additional to the Contract price of ten cents on each musket.*[84]

These changes were approved and thus for the first time an American copy of Blanc's 1777 musket lock, Fig. 5.84, is encountered and was to appear again on a variety of muskets and rifles through to 1830. It is further testimony to the enduring quality of Blanc's design.

When it comes to American lock making, again there is very little information. It has been suggested that this may be because most gunmakers did not make their own locks but bought them ready-made from foreign suppliers[85] or that the lock making process had changed so little that there was nothing new to record.[86]

Lockplate manufacture at Harper's Ferry followed a similar path to that described elsewhere: forging between dies, annealing, and pickling in dilute sulphuric acid to remove firescale, but unlike elsewhere, the flat surfaces were ground, not filed. It was then drilled using a drilling jig to correctly place all the holes, followed by tapping those which required threads.

Smith[87] provides two particularly interesting sidelights on lockplate production; one is that the lockplate had *projections that centered it in the drill jig*, and although he does not state what they are, they could only be the 'trapezoidal eminences' or 'bolsters' on the 1777 lock – this reflects practices at Tula; the second comment relates to the filing of the lockplate to shape, in that, *With the newly drilled holes serving as bearing points,* the lockplate was secured in the filing jig.

This latter small but important detail has not emerged until this point, however much it might have been expected, and may indicate the earlier, or an alternative, approach to this process. However, there seems little to suggest that there was any new approach to making the other lock components – sear, tumbler, bridle, cock, etc.

The first evidence for the use of 'hollow mills' for machining tumblers and screws in America was at Springfield Armory in 1809[88] but one is left to wonder whether they were copies of the 'English' tumbler mills.

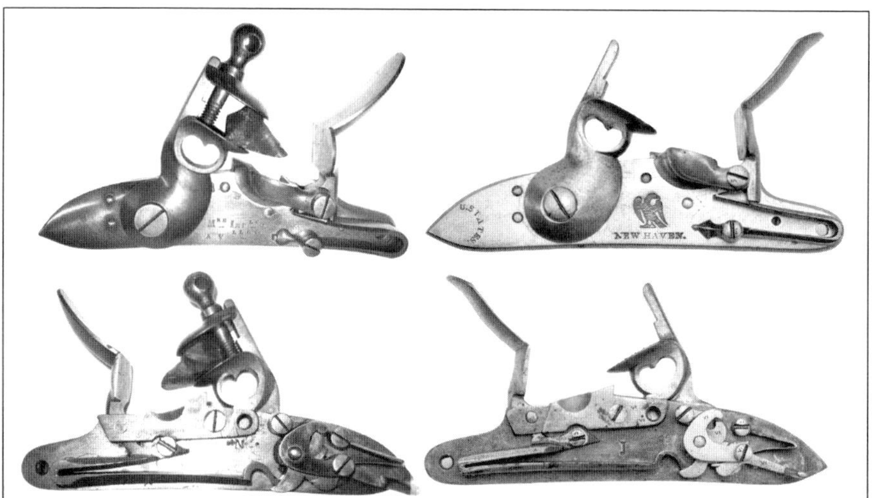

Fig. 5.82. Left: the 1777 lock. (XII.2231 © Royal Armouries); Right: Whitney's 'improved' 1798 contract lock (SPAR 4662 © Springfield Armory National Historic Site)

In 1813, Simeon North received a contract from the U.S. government to manufacture twenty thousand pistols, Fig. 5.83. What is especially interesting is that this contract specified that *the component parts of the pistols were to correspond so exactly that any part of one pistol may be fitted to any other pistol of the twenty thousand.*[89] This has been claimed as the earliest known contractual requirement for interchangeability but it has already been noted that the earliest known contract of this nature was issued by the French government 90 years earlier.[90]

North's pistol, like Whitney's musket, also embodied Blanc's 1777 style of lock with its characteristic sloping pan.

North's invention of the milling machine, Fig. 5.84, may have been an outcome of this venture. A claim by Roe[91] and propagated by Woodbury[92] and in The American Machinist,[93] was that Whitney invented the milling machine during his venture into gunmaking. A recent study, however, makes a strong case for it being the brainchild of Simeon North around 1816 (Battison, 1984).

The machine and cutter bear some similarity to the rotary file used for shaping the pan illustrated by Diderot.

In 1828 Simeon North was granted another contract with the U.S. Ordnance Department to supply five thousand Hall rifles. He built his own machines for the

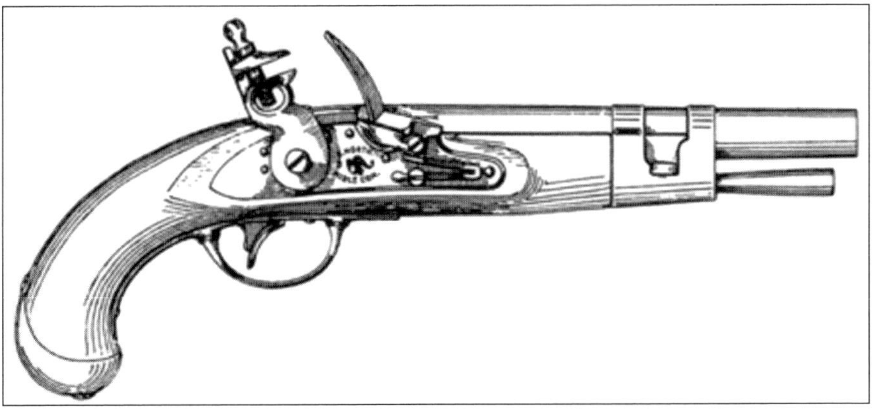

Fig. 5.83. Simeon North pistol M.1813/M.1816. (Hicks, 1957, p .20)

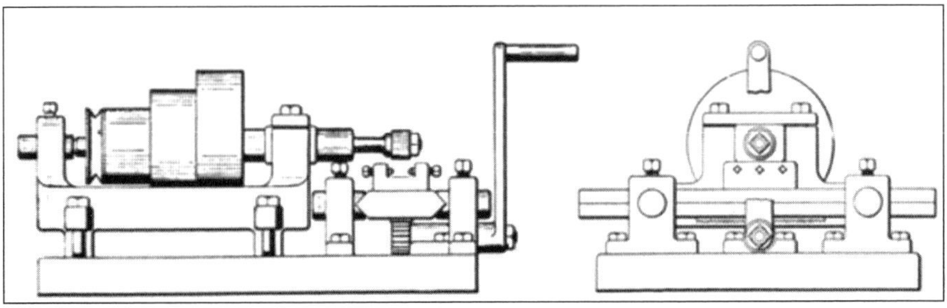

Fig. 5.84. Simeon North's milling machine. (Ashburn, 1977, p. B 12)

job – again, we do not know what they were like but some were probably copies or variants of his 1816 milling machine. By 1834, working to a set of gauges supplied by Hall, North was also producing interchangeable rifles. Not only that, the components of Hall's and North's rifles, produced at widely separated establishments, were interchangeable. This was a major achievement in the evolution of precision manufacturing and Smith[94] suggests *no two individuals played a more important role in this development, or as machine-tool innovators in general during the early 19th century than did North and Hall.* But, as Gordon has noted, the ability of craftsmen to achieve the manufacture of standardized, precision metal products in quantity, with work spread amongst several craftsmen, began at least as early as 1537 when Georg Hartman, a prolific instrument maker in Nuremberg, Germany, made large numbers of 'identical' astrolabes.[95] However, whether their precision matched the requirement of interchangeable gun locks is not known.

While no-one can deny the importance of North and Hall's achievements, Smith gives scant acknowledgement to Blanc's achievements or to those at Tula. But, while Smith was aware of Blanc's work, Gamel's original book is an extremely rare publication in the original Russian and it had not appeared in translated form at the time Smith was writing.

5.5 Interchangeability

Cotty[96] dismisses Blanc's claim of interchangeability as *a chimera*, and cites problems encountered with some of Blanc's methods, which Blanc himself had already recognised and overcome. By the time Cotty was writing, Blanc, Gribeauval, de Brienne, de Borda, and le Roy were all dead and so were unable to contest Cotty's views. Alder[97] states that none of the locks made by Blanc have survived; none have been encountered as being documented during many years of research but recently evidence has come to light which suggests that Blanc may have achieved interchangeability.

At Blanc's Roanne factory, he had, by 1800, shipped 11,500 interchangeable gun locks, in the un-hardened state, to other manufactories, including St Etienne, Liège, Turin and Versailles, where they were engraved or stamped with that manufactory's name and fitted to muskets.[98] In 1801, with Blanc close to death, the Roanne factory was bought by Jean-Francois Cablat and, until its closure in 1807, produced an annual average of 10,000 interchangeable locks.

This supply of un-marked locks taken in conjunction with the crudely marked Versailles lockplate and the absence of *assembly marks*, Fig. 5.85, strongly suggests that this lock is *interchangeable* and may be from Roanne. It is the only musket in Royal Armouries collections of 1777 muskets that has a lock of this nature.

The evidence provided by another lock, Fig. 5.86, also suggests interchangeability was being achieved in Liège. Considering Cotty's statement that at the time of his writing, the lock factory at Roanne had been transferred to Liège,[99] the possible interchangeable nature of this lock might not be so surprising, whether made in Liège or as an earlier import from Roanne!

To these can be added the Tula lock which on close examination is also seen to have no assembly marks, Fig. 5.87.

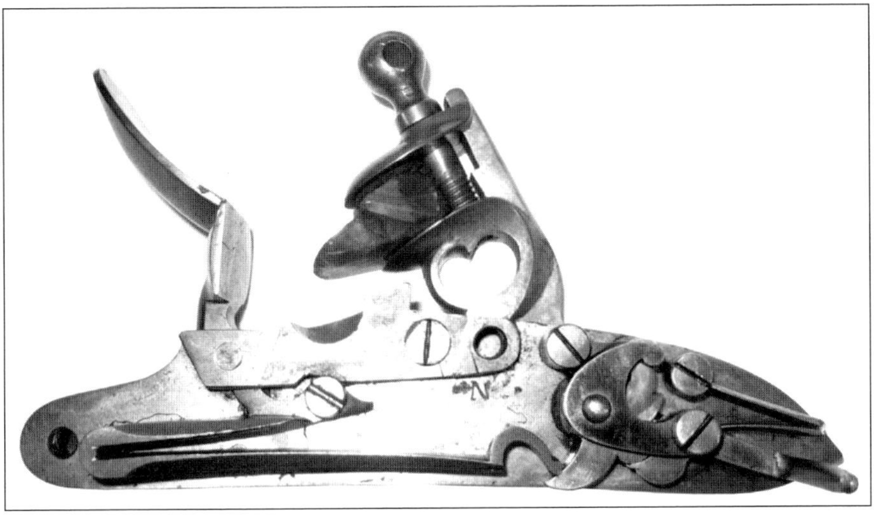

Fig. 5.85. The inside of the M.1777 lock marked 'Versailles' showing complete absence of the 'assembly marks' characteristic of non-interchangeable arms. (XII.2231, © Royal Armouries)

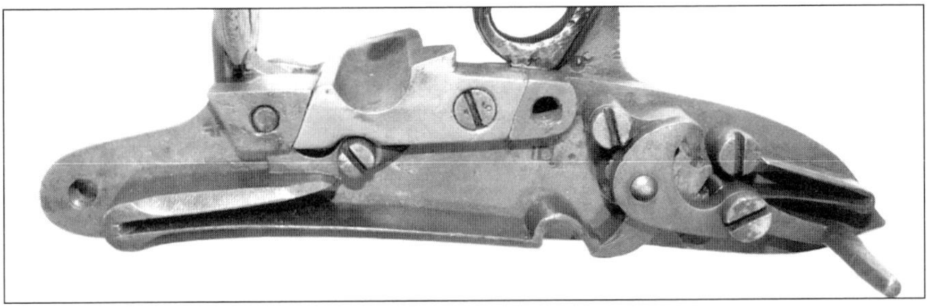

Fig. 5.86. Interior of the M.1777 lock made in Liège. (Author's collection)

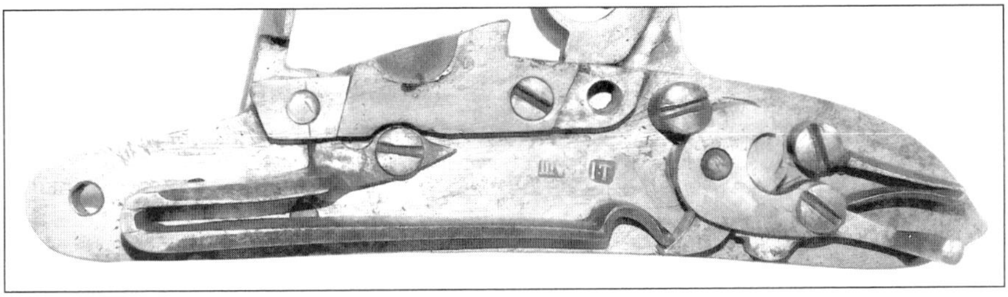

Fig. 5.87. Russian M.1808 lock made at Tula showing absence of assembly marks. (© Royal Armouries XII.2351)

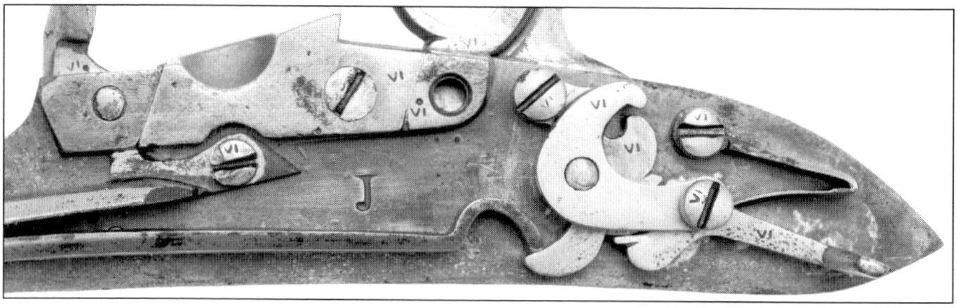

Fig. 5.88. Detail of the interior of Whitney's lock showing 'assembly marks' on all components. (SPAR 4662 © Springfield Armory National Historic Site)

This places it on a par with Blanc's locks from Roanne and the one from Liège in their lack of assembly marks, a factor which does not appear to have been recognised and reported upon before.

These locks, therefore, occupy not only an interesting, intriguing, and mysterious place in this study but also a potentially important place in the history of firearms manufacture as the earliest surviving examples of interchangeability. However, examination of a greater number of specimens is needed to be able to establish whether these are interchangeable within groups of similar items.

In contrast, close examination of the Whitney lock, Fig. 5.88, reveals a plethora of traditional 'assembly marks' on components which suggests that he did not achieve interchangeability, despite claims to the contrary.[100,101]

This lack of interchangeability is supported by the suggestion that America did not have sufficiently precise tooling at this early date[102] and an analysis of tool marks on excavated components from Whitney's factory site by Gordon,[103] suggests extensive use of files and little mechanisation.

Chapter 6

Manufacture of the Stock

6.1 Introduction

Whilst the barrel and lock, or other firing system, are the key items of any firearm, it is the stock which unites them and upon which are mounted all the parts of the gun in their correct juxtaposition. It is a vital part of that triumvirate, 'lock, stock and barrel'.

The writers of *Espingarda Perfeyta* recognise the requirements of a gunstock and the best wood to provide them:

> *There are various woods of which Stockmakers make the Stocks, but all must choose Walnut, for it is the best, and possesses those qualities which make it more proper for being worked, and also for its security, this being what makes it more durable. It is light to handle, attractive by nature, and lastly it keeps straight, a singular quality for our purposes, for these reasons taking preference over Cherry wood to which for a nice appearance new colours are given, and as they are artificial, it loses grace for it is not clean and has stains; over Maple wood which, since it is white, is not right. It is also better than Brazil wood, which is too heavy. Thus, only walnut should be chosen, and this from the heart of the timber where are to be found the good qualities we have described.* [1]

6.2 Stocking the French M.1777 musket

Writing almost a hundred years later, Cotty is of the same opinion regarding walnut as the best choice:

> *Walnut wood offers more advantages than others for the assembly of firearms, and is the only wood used for military firearms. This wood must have at least three years in storage after cutting, including two under cover. If a wood is not sufficiently dry, by giving it a stroke of a plane, the chip is still wet and it has the smell of green wood; exposed to the sun, the wood splits when it dries.*

> *These woods must be grey or half grey; the white ones almost always coming from the sapwood, which is an imperfect part, must be discarded, as well as those with yellow stains and worm holes, or other defects which show their decomposition or bad quality; finally, those who are not sufficiently straight in their whole length, or which have knots which can cause them to break, must also be discarded. In general, their quality depends on the terrain in which they grew, their age, and how they were treated after cutting. They can be kept forever by rubbing them at intervals with a piece of serge or cloth impregnated with oil.*[2]

It is noteworthy that Royal Armouries have in their collections seven French Model 1777 An. XIII muskets, items: XII.2228, XII.2229, XII.2230, XII.2231, XII.2232, XII.2233 and XII.2234, dated 1806, fitted with stocks of beech, poplar, maple, chestnut, ash, oak and elm. These were reportedly part of a series of experiments to find a suitable alternative to walnut as the forests of France were depleted of this timber.

Those views on the selection of the timber are equally valid today, but sadly, these two writers, who provide a wealth of detail on the making of the other parts of firearms, have little or nothing to say about making the stock.

However, this deficiency with regard to the M.1777 musket is more than made up for in Dale's publication of 1810, *Mémoire sur la mise en bois ou montage du fusil, la trempe de la platine, la visite et la recette de l'arme finie* [Memoir on the setting in wood or assembly of the musket, the hardening of the lock, the inspection and acceptance of the finished weapon]. It describes practice at the Imperial Arms Manufactory in Liège and has thoughtfully been reprinted under the guidance of Claude Gaier, former Director of the Museum of Arms in Liège, as part of a study of Belgium's industrial archaeology. It is apparent from Dale's tome that 'setting-up', to use English terminology, consists of far more than making the stock and fitting the various parts to it. A wide variety of skills – perhaps more so than any of the other craftsmen who were highly specialised in a narrow discipline – were needed. Setters-up had to have both woodworking and metalworking skills – it has already been shown that the vital hardening, tempering and, to some degree, the finishing of lock components fell to them. In addition, they also carried out similar work on the other metal fittings.

Since, as far as is known, no mechanisation was used in Europe until the mid-19th century, it was a very labour intensive, hand-craft process and it can be reasonably safely assumed that what Dale has to say about stock making in Liège is equally applicable to most, if not all, gunmaking at that time.

Selection of timber

Dale also notes that walnut was the wood used for the stock and must be without defects or damage which could affect its durability. It was expected to last fifty years but could last longer if carefully treated. He is of the opinion that it was impossible to estimate the time necessary for this wood to lose its strength if it was not submitted to the action of the elements.

The walnut tree grows slowly, and the best was regarded as one which had grown on ground a little elevated and sloping - those from lower ground, or exposed to floods, or to a soil with too much clay, were the poorest and subject to the most defects. Also, one which grew quickly had the most sapwood and gave, consequently, a smaller quantity of sound wood and it also had lower specific gravity, solidity, and durability than that which grew more slowly. It was to be felled when 60 to 100 years old or more but only gave good service when at least 70 years old. Its age could only be judged after felling by the annual rings which surround the heart of the tree. It was only to be cut in autumn, otherwise it would be subject to mould because of the sap it contained before its growing period had finished, indicated by the loss of the leaves.

Before use it had to be left for three years, including two years and at least eight months in a covered store. The trunk alone was suitable for stocks, and it was first roughly squared then sawn into planks 2 [Paris] inches and 2 to 3 lines thick ready for cutting to shape. This approximates to 60 mm and a finished stock measures around 46 mm at its widest part.[3]

Making the stock 'blanks'

This was done by employees of the contractor if he was not able to buy the wood already cut into blanks by the merchants who supplied it. The starting point was a plank 8, 10, 12 and more feet long. On each of them the foreman drew the outline of the stocks that could be obtained by placing several templates of thin wood and adjusting them to be able to fit as many as possible, being careful to avoid any defects which might have been encountered, such as knots, splits, rotted portions, etc. The blanks were then cut out by sawyers, each using a frame saw with a 2 feet 6 inches (760 mm) long blade placed centrally in a frame about three feet (914 mm) wide.

A plank 10 feet (3 metres) long and 18 inches (450 mm) wide, having an inch of sapwood, was sufficient for 10 musket stocks, of which it was probable that there would be at least one to discard before the work on it was finished due to faults not initially apparent. Each sawyer cut 25 to 30 stock blanks per day.

The blanks, before being accepted into the Manufactory's stores, were examined by one of the contractor's employees in the presence, usually, of one of the inspectors, to exclude any that had visible faults. Only then was it marked with the inspector's stamp. The causes for rejection, sometimes not encountered until work had begun, were:[4]

> *white wood* ... sapwood
> *cracks* ... splits discovered on the surface of the wood, arising from the action of the sun on the planks, especially if they were previously exposed to rain.
>
> *galls* ... damage in the body of the wood, such as blackish crevices, arising from blows received by the young tree, such as nails hammered in, etc.

knots ... in the parts to be reduced to small dimensions

If the wood was warm ... arising from having left planks stacked or overlapping one another and exposed to the rain; trapped water produced this effect, recognized by the water stains revealed by planing the wood at the relevant place.

green ... veins or spots of varying width with a bright yellow colour, showed the presence of worms especially in the sapwood

red-green wood ... was also discarded if it was worm-eaten.

marrow-hole ... was a circular hole, 2 to 3 lines (4.5 – 6.75mm) in diameter sometimes found in the heart of a tree which had reached a certain age - planks with this defect therefore gave fewer stocks than those from the edges.

Roughing-out

The wood, after having spent at least eight or ten months in a store, sawn and stacked in a grid, was then transported to a second store where, before being used, it remained for two years. During the early stages of this period, they were 'roughed-out' so that they remained in store in that condition for as long as possible to allow them to acclimatise before being finished. In addition to various tools, the person roughing-out was supplied with a wooden template or pattern for the profile, a measure for the thickness and a butt plate.

The roughing began by placing the rough stock in a vice and planing it all over to reveal any defects and remove them if possible, without reduction below the dimensions required. A template was then applied, the outline drawn, followed by removing the excess wood with the plane.

The excess wood at the heel of the butt was sawn off, followed by drilling the recess intended to receive the plug on which the mark of the inspector in charge of examining finished weapons was ultimately placed, Fig. 6.1.

The profile of the underside was then marked, followed by forming the comb. The small [wrist] of the stock was taken to its rough dimensions by use of the chisel, the profile above the small of the stock where it meets the comb was marked and the comb cut to its correct length and finally planed it to its correct thickness. The craftsman then put his mark on it. At this point the wood is about 4 lines (10mm) oversize in all dimensions, except where the mark is, than in the finished musket.

Fig. 6.1. The stock inspection and final inspection stamp after completion of the work on and surrounding an inserted wooden plug on the butt. (XII.2230, © Royal Armouries)

Each worker 'roughed out' about 30 stocks a day in the summer and 22 in the winter. One could be prepared in about 15 minutes but this tiring work required quite frequent rest.[5]

Amongst the rough stocks were some that had faults which made them unfit for service, sometimes due to clumsiness of the rough-stocker but more often because these defects were deep in the thickness of the wood. In either case, they were rejected and the work on them not paid for.

As one example, out of 14,400 stocks roughed out, 1317 were rejected, which was 9 to 10 per cent, an average quantity, even after careful selection before entering the manufacturer's stores.[6]

Mounting the components – "setting-up"

Sets of stock blanks, barrels, locks, furniture, ramrods and bayonets were issued to each craftsman and the final assembly of the musket could only be undertaken after a variety of tasks on both the stock and its metal components had been completed, and these had to be arranged in such a way that no time was wasted. It was found that the most advantageous arrangement for a workshop was one comprising a master, two assistants and an apprentice, who, fully occupied, could complete four muskets in a 12-hour day.

The stock, albeit being basically a carved piece of wood, was unlike any other piece of carved wood, so setting-up required an extensive range of tools. These are listed by Dale, but he does not illustrate them. However, their natures will become apparent from descriptions of their use and the tasks they performed.

Several jobs had to be performed on the metal components and these were interspersed with work on the stock and the finishing, hardening, tempering and polishing of lock components has been covered in Chapter 5. Polishing the furniture, bayonet, ramrod and barrel was done by the apprentices. It was often found that these processes revealed some fault with a component. The setter-up had to inspect them carefully and report to the factory any defects to have them examined. Those found to be faulty were marked as scrap if they could not be easily repaired, and the setter-up received replacements.

Likewise, the setter-up reported any faults he discovered in a stock he was working on, being careful to preserve the mark of the inspector of roughed-out stocks to show that a fault in the stock had not been caused by the setter-up. Also, when the musket was close to being finished, before he removed this mark he had it struck again so he could finish the stock.

The breech tang screw had to be threaded, finished and case hardened; the hole in the trigger plate had to be drilled and tapped to receive the barrel tang screw; the heel of the breech screw needed to have the notch formed to allow the passage of the rear lockplate screw; the trigger was drilled for its suspension pin; the mortices in the trigger plate, the trigger and the swivel tenon were formed along with that for the hook of the trigger guard; the mortice in the trigger guard for the swivel mortice was formed to match that in the trigger plate.

Any case hardening was arranged in such a way that the setter-up would always have time to exchange any faulty piece and harden replacements if necessary.

Commencing work on the stock

The 'roughed-out' stock, which having been felled three years previously and having spent two years and eight months in store, was very dry and no longer subject to breaking or warping, so was in the most suitable condition for setting-up. The setter-up had a workbench furnished with an iron vice and a wooden support, like that used in the roughing-out process, and had various tools and items of equipment including a template, in thin wood, for the profile; a gauge for the top slope, up to the rear of the middle band so its position could be marked; a gauge for checking the lower curvature of the butt, the width of the wrist and the 'reinforced part' where the lock was fitted; a gauge for determining the position of the middle band on the infantry and dragoon muskets. [The location of the top band was determined by the socket of the bayonet, the lower end of which, when fitted, had to be flush with the foremost end of the band.]

The butt

The setter-up began by tracing on the roughed-out stock the profile of his template. The rear face of the butt was shaped with a plane and a rasp, followed by shaping the top of the comb at the heel of the butt where the tang of the buttplate would fit. Having traced its contour, a 'V' hooked gouge was used to clearly outline the recess. This was then formed with the chisel to the required depth, determined by the thickness of the buttplate tang. The butt plate was tried on the wood and adjustments made so it fitted firmly on its housing, Fig. 6.3.

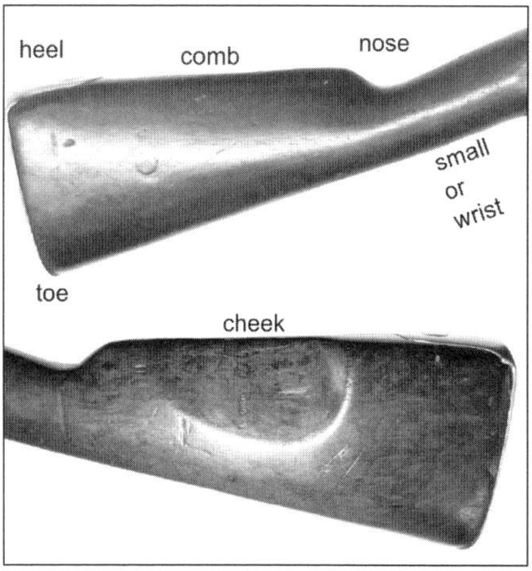

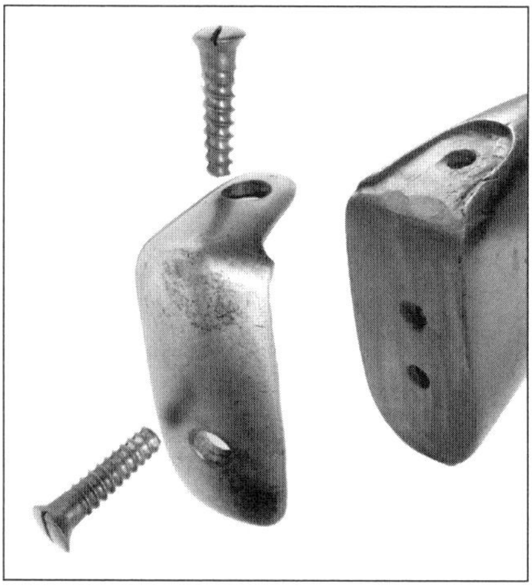

Above left: Fig. 6.2. The butt and the terms for its features. (© P. Smithurst)

Above right: Fig. 6.3. Finished seating with its butt plate. (© P. Smithurst)

Bedding the barrel

The roughly formed fore-end was sawn along its length, following a line representing the horizontal centre line of the barrel. The cut ended at the place where the breech end of the barrel would be and was then sawn through crosswise to create a step. The long sawn face was then straightened and smoothed with a plane.

The next stage was to place the barrel on the prepared surface, ensuring it lay on the centre line, and trace its outline with a scriber, Fig. 6.4.

Having traced the outline, it was then enlarged using a pointed hooked gouge, followed by starting the channel at the chamber end with a small chisel and continuing along its full length with a gouge and mallet. It was then finished roughly with a barrel rebating plane having a semi-circular convex face. The recess for heel of the breech plug was then traced and cut out with chisels. The barrel was replaced and its outline traced more accurately, along with the outline of the breech plug tang, which was prepared with a small gouge and chisel, Fig. 6.5.

The barrel channel was then finished to a semi-circular profile using a gouge and the barrel rebate plane. To check for size, the underside of the barrel was smeared with oil and placed in the channel to reveal any high spots. These were removed and this process repeated until the barrel, pressed against the wood in a vice, was deemed to be a correct fit.

Bedding the lock

The wrist ends at the widest part of the stock at the breech end of the barrel and which, on its side faces, accommodates the lock and, on the opposite side, the side-plate.

Dale (1810, p. 24) states that the lockplate complete with the pan, the bolsters and the inner edge of the pan being flush with each other and having been smeared with oil, was placed on the stock to leave an imprint of these features, Fig. 6.6.

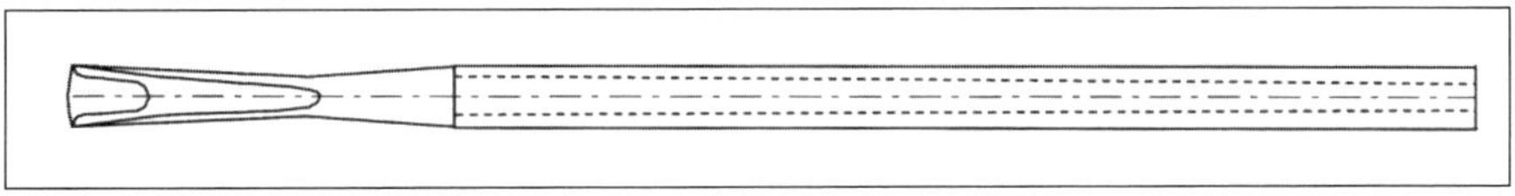

Fig. 6.4. Sketch showing the outline of the barrel on the fore end. (© P. Smithurst)

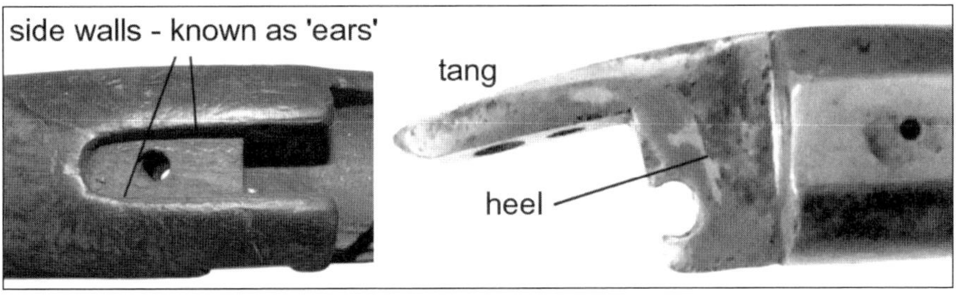

Fig. 6.5. Recesses for the breech plug tang and heel prepared. (© P. Smithurst)

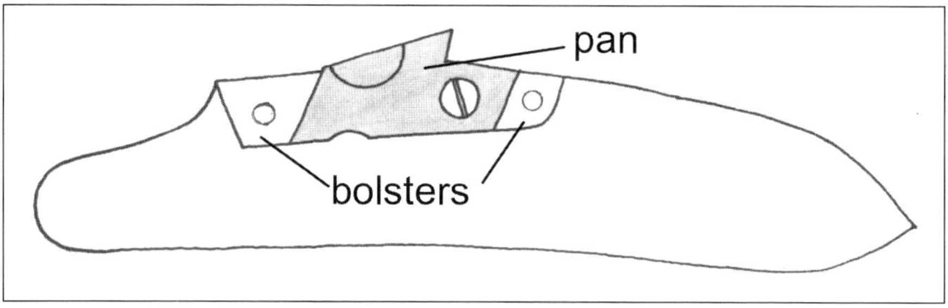

Fig. 6.6. the inner face of the lockplate with pan in position. (© P. Smithurst)

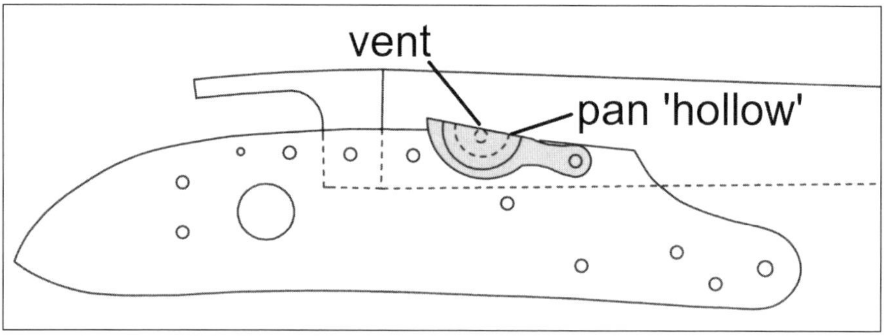

Fig. 6.7. Correct alignment of lock and barrel. (© P. Smithurst)

Dale notes in his description of the inspection process: *the vent must be in the middle of the width of the hollow of the pan and a little below the straight line connecting its side*[7] and this could only be accomplished with the vented barrel in place to ensure that the lock was 'bedded' in its correct position, Fig. 6.7.

Having marked the correct position of the bolsters, the wood was removed using various gouges followed by placing the lockplate again on the stock and tracing its outline with a scriber. The recess for the lockplate was then cut out to a depth equal to the thickness of the lockplate. The recess for the bolsters would also have been deepened during this operation and then any adjustments made to the depth of the lockplate recess by trimming its edges with the lockplate in place. At this point the positions of the holes for the two long lockplate screws – 'side-nails' in English gunmaking terminology – were marked and drilled with a brace and bit.

Having secured the position of the lockplate, space now had to be created for the internal components of the lock which presented a complex 3-dimensional geometry, Fig. 6.8.

Bearing in mind some of these components underwent movement when the lock was operated – the nose of the tumbler on which the tip of the mainspring rested and, when cocked and fired, the long branch of the mainsprings wept out an arc; the sear itself and its tang underwent similar, but smaller, movements when operated - non-moving parts had to be inlet to their exact depth, notably all the screw heads. All this had to be accomplished by removing the minimal amount of wood so as not to weaken the stock at this crucial point.

It was only after bedding the lock that another important operation could be carried out – filing the notch on the barrel breech plug heel to allow the passage

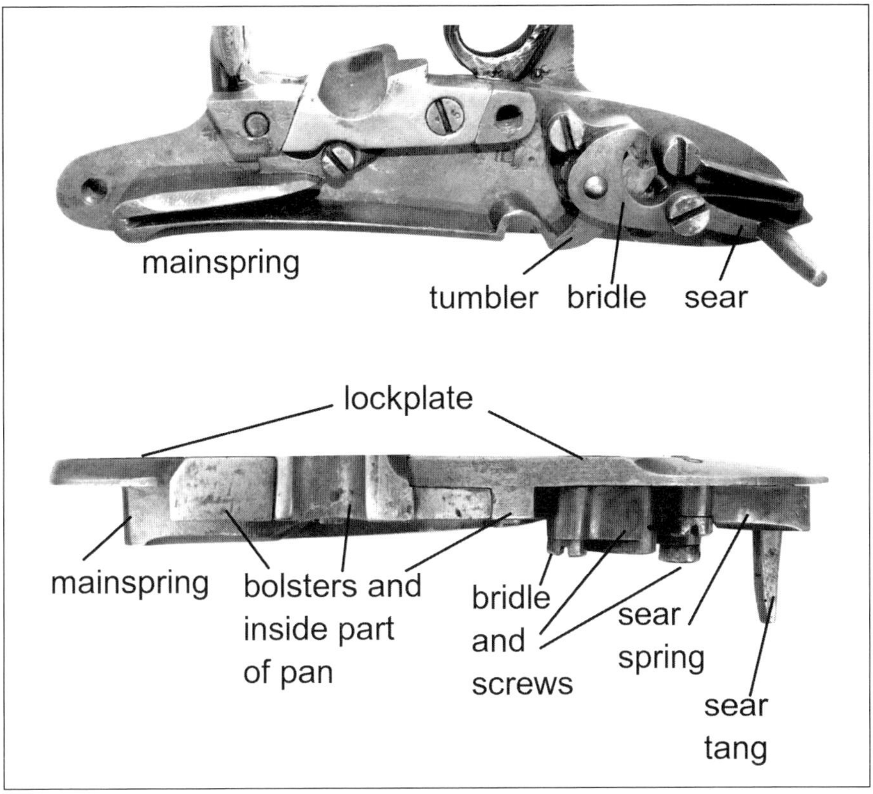

Fig. 6.8. Internal geometry of the lock. (© P. Smithurst)

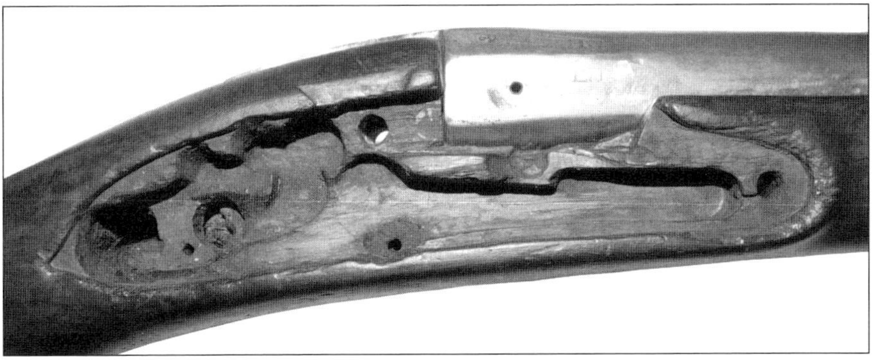

Fig. 6.9. The finished lock recess. (© P. Smithurst)

of the rear lock screw and, again, it was simply a matter of fitting the barrel and scribing the position of the screw on the heel, Fig. 6.10.

Having bedded the lock and drilled the holes for the lock screws, the opposite side was prepared for fitting the sideplate by ensuring the stock was at equal width on each side of the barrel and then the sideplate was in-let. The two lock screws were passed through the stock from the lock side, their positions on the sideplate marked, and the holes in the sideplate drilled. The sideplate could then be located using these screws and its outline traced on the stock and the inletting undertaken, Fig. 6.11.

The setter-up also forged the two large lock retaining screws - 'side nails' - and the breech screw. The shanks were formed with a hammer, then filed along with the heads. The ends were threaded with a die held in a stock fitted with a handle and the slot cut in the heads with a slitting file. The shanks were then rounded in a 'screw mill' using a hand brace fitted with a screwdriver bit, followed by making the underside of the head flat and square with the shank using a 'grinder'. The use of the term 'grinder' for a cutting tool with teeth is strange to say the least but the term was used in British military equipment terminology well into the 1860's,[8] Fig. 6.12.

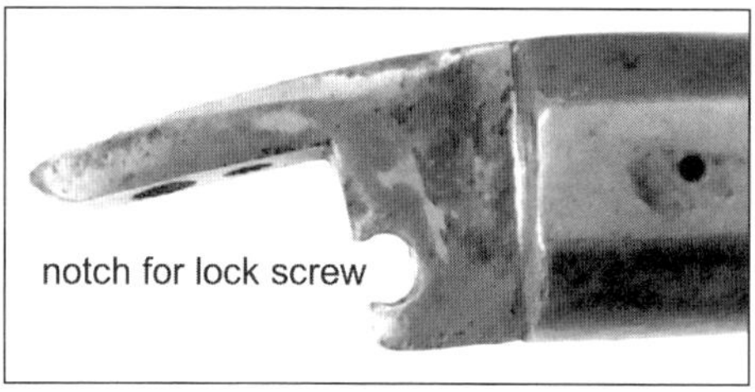

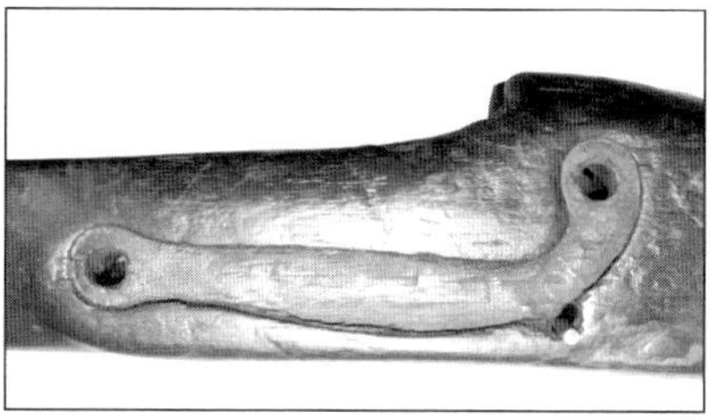

Above left: Fig. 6.10. notch in heel for passage of lock screw. (© P. Smithurst)

Above right: Fig. 6.11. Finished sideplate recess.

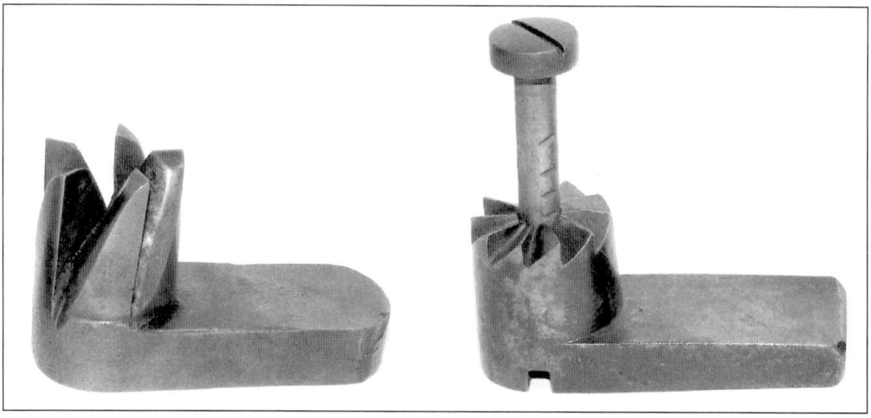

Fig. 6.12. Left: a 'grinder' for underside of head of breech tang screw; right: a grinder for finishing a shank and underside of the head of a 'side nail'. (Author's collection)

The projections allowed the tool to be held in a vice so the screw was horizontal and could be rotated using the hand brace fitted with a screwdriver bit in the usual way.

After bedding the lock and sideplate, work on the butt was continued. Using a draw knife, the borders of the lock housing were formed, sloping them at the front to meet the fore end, and maintaining the widths of the borders to within 2 lignes.

The 'wrist' was then rounded with rough and smooth flat files close to its correct size and shape and blending it in to the tail of the lockplate housing. The nose of the comb was blended into the wrist using a rough, half-round file, followed by smoothing the butt and the wrist with flat and half-round bastard files. The bottom edge of the butt and the wrist were rounded using a jack plane, draw knife and file. Major work on the butt ended by cutting out the cheek piece with a spokeshave or draw knife.

Commencing work on the ramrod channel

It should be noted that apart from bedding the barrel, no other work on the fore-end had been carried out until this point. Using a marking gauge, the depth of the barrel channel was scribed along the length of the outside of the fore-end and the edges of the ramrod channel, accurately centred, were then scribed along the length up to the point where the front of the bottom band would rest. The fore-end was roughly shaped on its underside with a spokeshave or draw knife and the channel partly cut using a small gouge and a special rebating plane.

Fitting the bands

The fore-end also had to have seating points prepared for the three bands holding the barrel tightly in place, along with recesses for the band springs to keep the bands in place. Bands first appeared on French military longarms with the Model 1717, only one band being fitted, the ramrod being retained by 'pipes' and the barrel by cross pins. The Model 1728 had three bands, including the familiar top band – the only one to be held by a spring – accompanied it would seem by the disappearance of cross pins. With the Model 1754 the top and middle bands were held by springs and the Model 1763 had all three bands fitted with springs. The Model 1777 initially

Fig. 6.13. The lower, middle, and top barrel bands. (Author's collection)

only had two springs but by 1800/1801 (Revolutionary calendar year IX) three springs were again being used.

Work began by using a small saw to create the shoulder for the bottom band, followed by using a chisel to remove wood to reveal this shoulder. The remainder of the fore-end was then shaped with a draw knife, followed by a hollow moulding plane and then a coarse flat file so that the band was a snug fit. Then, the shoulder of the middle band was created, followed by shaping the fore-end in the same way which allowed the top band to be fitted after any necessary adjustments. Its position was determined by its front end being sufficiently rearward of the muzzle to allow the bayonet socket to be fitted. The part of the fore-end between the bottom band and the lock housing was then rounded using a draw knife, jack plane and files.

Fitting the band springs

Although Dale notes that this is carried out later it seems appropriate to include the process here. It commenced with the top band which, unlike the other two bands, had no shoulder. With the top band in place, the spring could be laid alongside it so that its shoulder, Fig. 6.14, was a fraction below the lower edge of the band and the position of the root of the stock pin marked on the stock.

The hole for the stock pin was then drilled, the pin inserted, and the spring aligned parallel with the edge of the barrel channel, followed by tracing its outline with a scriber. It was then cut out with a mortice chisel and the edges cleaned with a chisel. After inserting the and spring, the top band was fitted with its bottom edge *under* the spring so the position of the retaining pin could be marked on the band and its hole drilled.

The springs for the lower and middle bands, Fig. 6.15, were fitted in the same way but reversed so that the shoulder on the spring held the band against its shoulder on the stock.

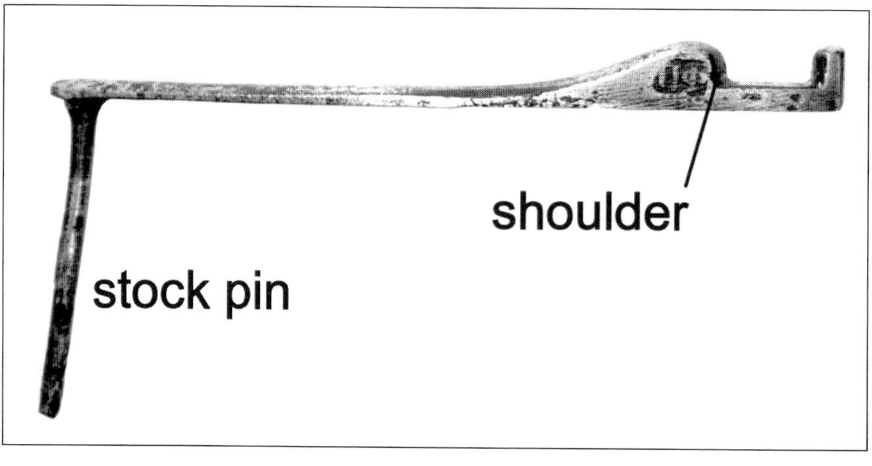

Fig. 6.14. The top band-spring.

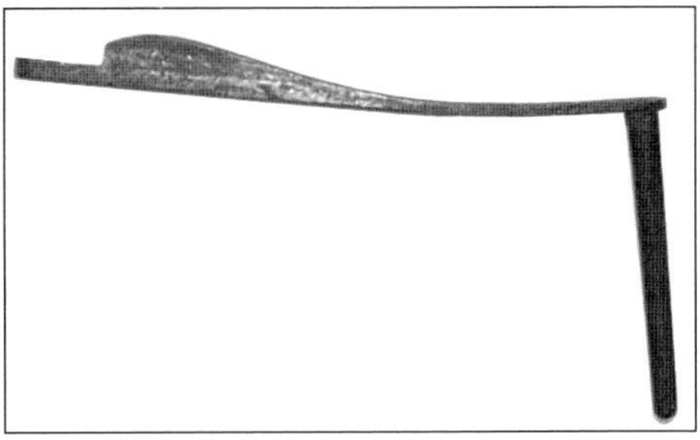

Right: Fig. 6.15. Spring for the middle and bottom bands. (Author's collection)

Below: Fig. 6.16. Seatings for the lower and middle bands and the bands fitted. (Author's collection)

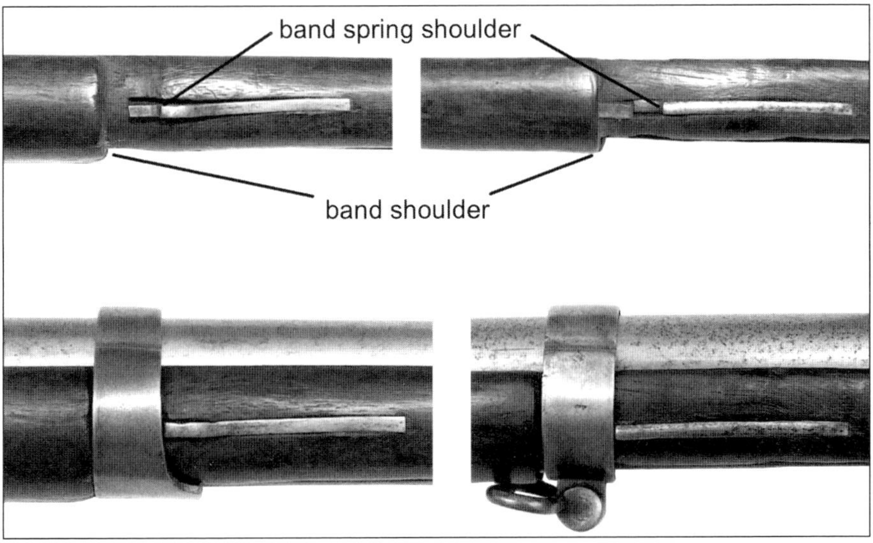

Completing the ramrod channel

Before discussing this it is worth noting that the ramrod was held in the stock by a spring set into the bottom of the barrel channel between the bottom band and the lock. The ramrod channel is self-explanatory but strictly speaking it is only a channel from muzzle to the bottom band, at which point it becomes a hole bored through to meet the ramrod stop on the trigger plate. The ramrod spring was used to hold the ramrod in place but Dale has little to say about its fitting although it is clear from Dale's narrative that this at least partly took place before the ramrod hole was drilled; the spring and its finished housing are shown in Fig. 6.17.

The drilling used a long drill in a hand brace and when the hole for the tip of the ramrod spring was reached, it allowed a judgement to be made as to whether the drill had followed the correct course. If not, it could be redirected by pushing it to one side through the hole for the spring. If all was on course, drilling continued to a point within about an inch of where the tenon of the sling swivel on the trigger plate would lie. The ramrod channel was not completed until after the trigger plate

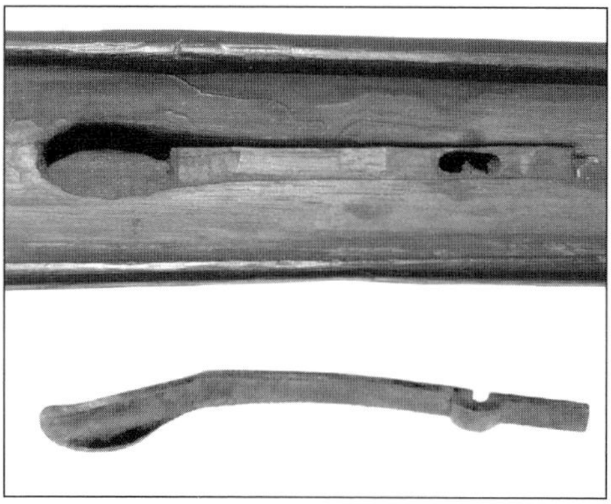

Fig. 6.17. The ramrod spring and its housing in the fore-end of the stock. The spring acted like a finger with a concave tip bearing down on the ramrod shaft.

had been fitted. Drilling was then continued up to the hole for the tenon of the swivel, again judging whether the ramrod channel was properly centred. A second bit, slightly larger, was then used to enlarge the hole and adjust its position if necessary. The ramrod channel was finished using a special grooving plane, the hole reamed, and its mouth slightly funnelled using a gouge.

Fitting the ramrod spring

The nature of this spring has already been shown in Fig. 6.17 and its housing was completed using various gouges and chisels. It also needed a hole drilling for a cross pin which passed through the 'U'-shaped notch at the right-hand end of the spring shown in Fig. 6.17. This hole needed to be accurately placed so that when the cross pin was fitted it bore down on the bottom of this notch, fixing the spring securely and enabled it to function without it being loose.

Fitting the trigger plate, trigger guard and trigger

As can seen from Fig. 6.18, these and other associated components form an integrated assembly which fulfils a variety of functions.

The trigger plate was placed in position on the stock and the positions of the ramrod stop and the 'pyramidal bolster' – a reinforced portion into which the breech tang screw fitted – were marked. Clearly, these had to be on the centre line of the underside of the stock. The recesses for these parts were formed with a gouge and a chisel. The trigger plate was fitted into these and its outline scored with a scriber, and the forming of this recess began by using the hooked gouge followed by the use of a mortice chisel. The straight edges were cleaned with a chisel and the rounded ends with a gouge so that the trigger plate was a perfect fit. The recesses for the breech screw bolster and ramrod stop were then finished, the bottom of the trigger plate recess trimmed to make the plate sit evenly along its length and the small recess for the trigger guard's rear securing hook created.

The position of the hole for the breech screw was marked and drilled half-way and then drilling completed from the trigger plate side. It can be seen that the breech tang and trigger plate are roughly parallel, Fig. 6.19, so this drilling could be easily accomplished by an experienced craftsman. The hole in the trigger plate was tapped, the breech screw fitted, and its threaded end filed flush with the trigger plate.

Dale does not pass any comment, but it is assumed that the slots for the trigger and the tenon of the trigger guard swivel would be cast in the trigger plate and would need dressing with files. They could then be used to mark their positions on the stock and the necessary mortices cut out with a small mortice chisel. Having ensured that the trigger moved freely without too much sideways play in the

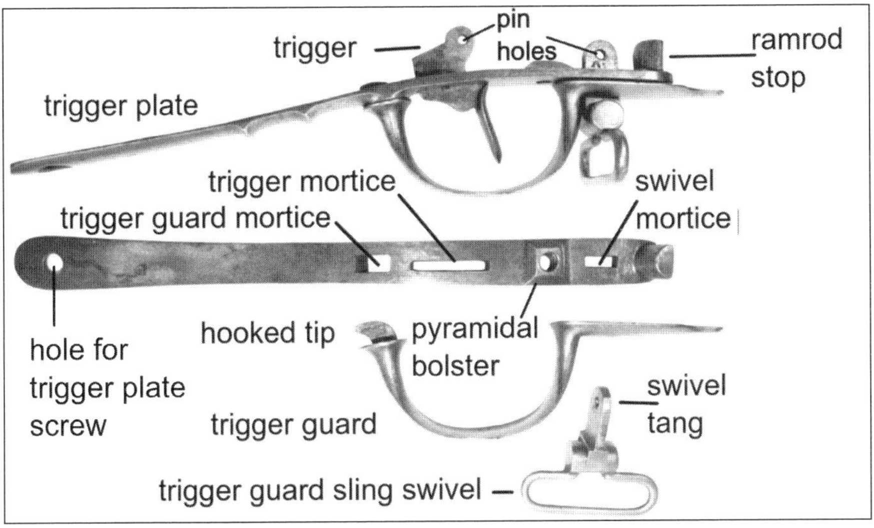

Fig. 6.18. Trigger plate and guard assembly. (Author's collection)

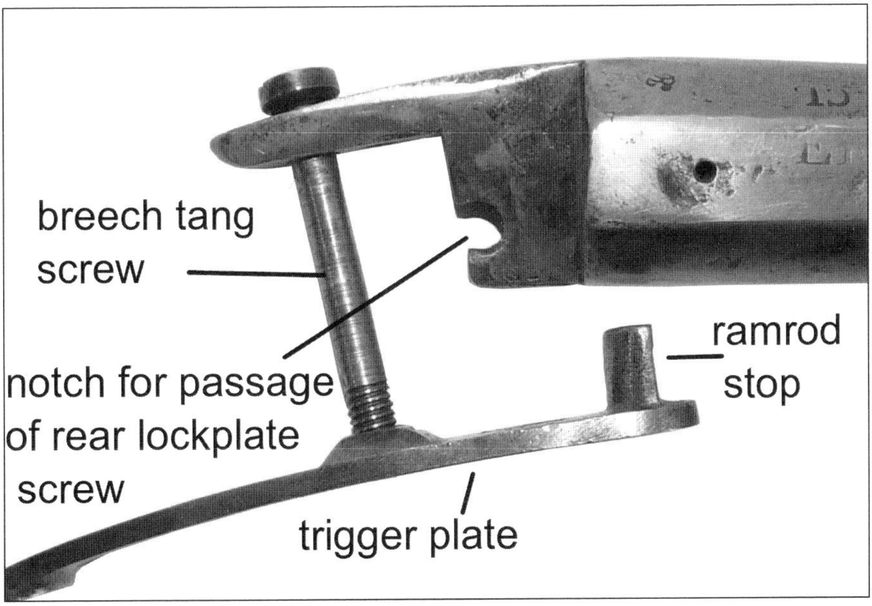

Fig. 6.19. Relationship of barrel tang, tang screw and trigger plate. (© P. Smithurst)

mortices of trigger plate and stock, the plate with the trigger in place was offered up to the stock so that the trigger was in contact with the tang of the sear, and any adjustments made to the rear of the trigger 'blade' in contact with the sear to ensure a correct fit with only minimal fore and aft movement of the trigger in its slot with the lock in its 'fired' or 'rest' position. It was important that when the trigger was fitted, it did not place any pressure on the tang of the sear until the trigger was pulled, otherwise it would have rendered the gun unsafe when fully cocked. The position of hole for the trigger pin could then be marked on the stock and drilled, followed by re-fitting the trigger, marking the position of the pin-hole on it, and drilling the hole with a bow drill. A similar procedure was used to fit the tenon of the swivel fitted in front of the trigger guard.

Finishing the stock

The stock was now ready to be given its finishing touches. The fore-end and wrist were smoothed with files of various grades. These parts, along with the butt, were then scraped with a scraper, the cheek being ignored at this stage. The stock was then rubbed all over with a skin on which fine, sharp sand was glued with a strong glue. To maintain sharp edges, the cheek was then finished with a draw knife and a smooth half-round file, followed by using the same file to finish the borders surrounding the lock.

After this final work, the setter-up stamped his mark on the stock and then assembled the finished musket and presented it for inspection and acceptance into store.

Dale noted that the total time for all the work in setting-up one musket was 12 hours.

6.3 Stocking the Tula M.1808 musket

It is clear from Gamel's account that stock making and setting-up at Tula was carried out on closely similar lines so only the differences will be discussed here. These were principally in the use of elm or birch in place of walnut, some of the tools employed, and possibly even the place of work which was usually the stockmakers' own homes.[9] Also, the furniture was of brass. They were supplied with all the components to build a finished musket.

Bedding the lock

The lock was laid against the stock with the barrel in place so that the pan of the lock was adjacent to the vent in the barrel, as noted by Dale, but then the process differs considerably. The position for the sear tang was marked, followed by drilling a hole to accommodate it. Only then was the lock bedded by carving out the recess using a variety of knives, chisels and drills. The holes for the screws – 'side-nails' – securing the lock in the stock were also marked and drilled at this stage. After fitting the lock, the excess wood around it was removed using a flat chisel.

The barrel and lock were then removed, and the position of the hole for the suspension pin of the trigger was marked and drilled, followed by cutting the slot which allowed the body of the trigger to make contact with the sear and operate it.

Finishing the stock

All the parts of the stock were adjusted to the reference dimensions by the use of planes followed by smoothing with wood files. Finally, the stock was scraped and polished with an unspecified resin. In order to give it the required colour, two materials are stated to have been used, although this, again, may be due to mistranslation! One of these materials, described as 'gun-stock liquor', was said to be nitric acid![10] The other is described as 'gun-stock-vodka', a strong vodka to which several small pieces of iron are added.[11] Iron salts, which the first mixture would provide, are known as stains for pale timbers and especially those which contain polyphenols, such as birch and elm, with which iron salts react (cf inks made from oak galls) to create darker substrates. On the other hand, the use of alcohol seems very unlikely since it would not have the same effect; although the water content might cause the iron to rust, it would not take it into solution. This has to remain a mystery at this time.

After painting with this 'liquor', the stock is dried over a fire and then rubbed with hempseed oil which, due to its ability to polymerise to form a resin, acts as a varnish and protects the timber.

6.4 Stocking in Britain and America in the early 19th century

In a footnote to his section on gun-stock manufacture, Gamel notes:

> *By viewing in Portsmouth the excellent machines invented by Mr. Brunell* [sic] *for the manufacture of ship blocks, it is easy to agree that mechanical methods could also be convenient for the manufacture of the gun-stocks.*[12]

These machines were certainly ground-breaking in many respects but the only one that seems to offer any scope for adaptation to the making of a gun stock is that used for the production of the 'dead-eyes', the circular blocks by means of which the standing rigging was secured to the sides of the ship. Strangely, their manufacture does not appear to be covered in the patent but they are usually included with illustrations of Brunel's and Maudslay's block making machinery at Portsmouth.

While dead-eyes were circular, they had two grooves which did not extend to the full circumference of their periphery and tapered off at their ends. The machine for cutting this groove used an ingenious mechanism in which a twin lobed cam attached to the spindle on which the dead-eye was mounted, operated on a *follower*. This was attached to a weighted or spring-loaded frame carrying a rotating circular

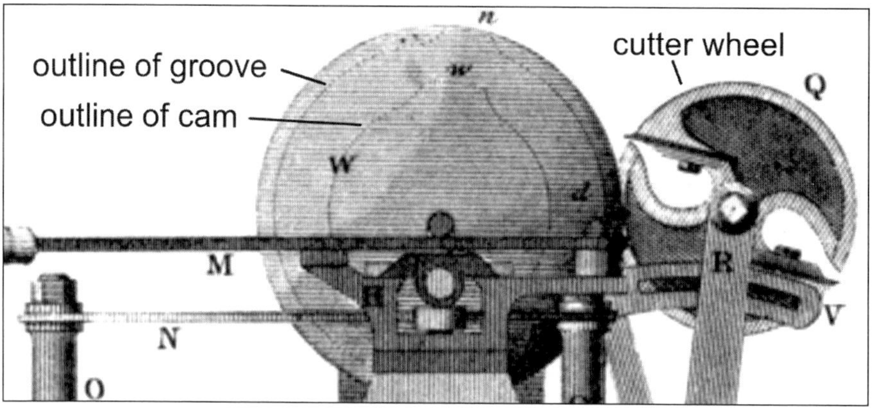

Fig. 6.20. Annotated detail to show the outline of the groove following the outline of the cam. (Rees' Encyclopaedia c. 1811, author's collection, © P. Smithurst)

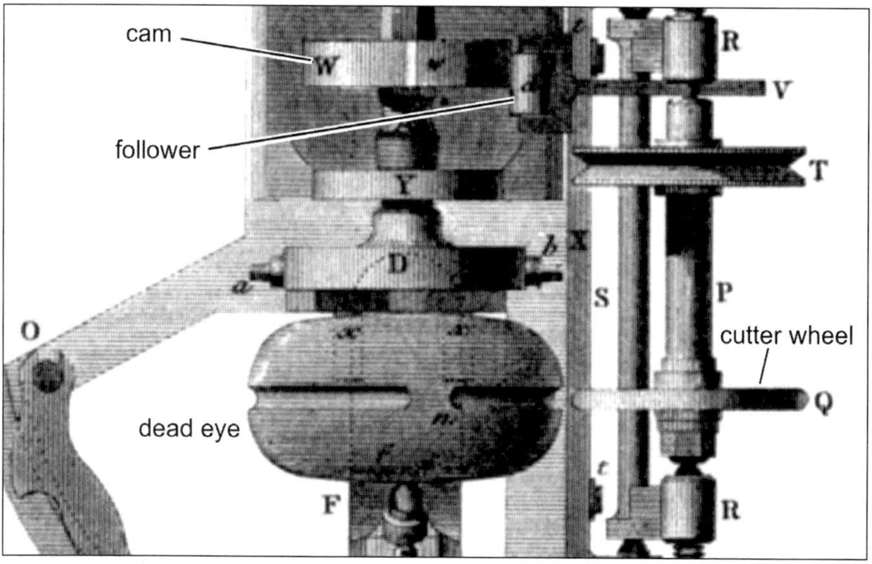

Fig. 6.21. Annotated engraving showing plan view of Brunel's machine for making dead-eyes. (Rees' Encyclopaedia c. 1811, author's collection, © P. Smithurst)

cutter, causing it to rock backwards and forwards and bring the cutter into or out of engagement with the workpiece. The arrangement is shown in Figs. 6.20 and 6.21 in which the broken lines delineating the cam and the groove in the dead-eye are more clearly visible.

Chapter 7

Conclusions

It is abundantly clear that gun making of any form was both an extremely labour-intensive process demanding high levels of skills in each of the multitude of operations which had to be performed. It is also clear that in regard to military firearms particularly, attempts were being made in some quarters to introduce standardisation and, beginning in France, that process began to be pursued to its logical extreme of 'interchangeability' with components of standardised sizes verified by a comprehensive system of gauges.

It was a period also in which we begin to see the evolution of machines to not only replace manual labour but to aid the achievement of uniformity. One of the final products to become 'machine-made' was the stock, which was undoubtedly one of the most complex wooden items in existence and required precision in its manufacture so that all components to be mounted on it were in exactly the right place. No-one, apart from Gamel, seems to have given thought to the adaptation of Brunel's machines to making gunstocks, although the cutter would need to move longitudinally as well as radially for the machining of a long item to be accomplished. This was a contribution made by Blanchard in the United States in 1819 and the set of machines he developed for carrying out most if not all operations on the stock is discussed fully in Volume 2 covering the manufacture of the Enfield Rifle of 1853. A British patent, No. 2652, granted to Joseph Bramah in 1802 might seem, at first sight of the title, 'Machinery for forming gunstocks *&c.*' to indicate progress but it fails to live up to expectation. It is primarily concerned with producing flat, linear, or spherical surfaces, and no record of it ever having been put into practice in any context has been located.

Appendix 1

Weights and Measures

Various weights and measures have been touched upon and it seemed helpful to create this appendix to explain some archaic measurements encountered.

English measures

The 'line' has, at various dates, been used as a measure of length by a number of countries but with varying degrees of discrepancy.

In England it was taken as $^{1}/_{12}$th of an inch, 0.0833" / 2.1 mm.

The 'finger' – the Roman 'digitus', was the width of a finger and taken to be ¾ inch / 19 mm. It also gave rise to the English 'nail', being the width of the thumbnail (Klein, Herbert A., 'The Science of Measurement', Dover Publications, 1988.)

Most common imperial measures:

1 inch (abbreviated as 1" or 1 in.) = 25.4 mm
12 inches = 1 foot (abbreviated as ft.) = 304.8 mm
3 feet = 1 yard (abbreviated as yd.) = 914.4 mm

1 pound (lb) = 16 ounces (oz) = 453.44 grams
1 ounce = 28.43 grams

In ballistics terminology, traditionally the avoirdupois system of weights was used:

1 ounce = 16 drachms = 28.43 grams
1 drachm = $27^{11}/_{32}$ grains = 1.78 grams
1 grain = 0.065 grams

Archaic French measures (Kline p. 72)

1 point (p)		= 0.188 mm
1 ligne (l)	= 12 points	= 2.256 mm

1 pouce (Paris inch) = 12 lignes = 27.07 mm = 1.0657 Imperial inches
1 pied (foot) = 12 Paris inches
1 toise (fathom) = six pied
1 aune = 120 cm

1 livre (pound) = 500 g = 16 'once'
1 once = 31.25 grams = 8 gros.

Portuguese Measures

Couto	half palm - across palm of hand without thumb. (*Dictionary of Real* [Royal] *Academy of Spanish Language*)
Hand	taken across the palm including the thumb and is approximately 4 inches/100 mm
Palm	distance between tips of thumb and little finger on outstretched hand, what in England would be the 'span', and taken to be 96.36 lignes or 217.4 mm or 8.64 inches. (Palaiseau, *Métrologie universelle, ancienne et moderne,* Chez Lavigne Jeune, Bordeaux, 1816, p. 109)
Finger	taken as being 3/4 inch / 19 mm (see under English measures)
Vara	1.1 metres, approximately 43.3 inches (Palaiseau, 1816, p.197).
Grain	approximately 0.11 inches
Tostão	'testoon' – a small Portuguese silver coin approximately 18 mm diameter; with a silver density of 10.5 grams/c.c., this approximates to a thickness of 0.75 mm.

Adarme	1/8 of an ounce, the ounce approximating to the English ounce. (Vieyra, 1827) (The Spanish adarme is $^1/_{16}$ ounce according to Clark, 1878, p. 177)
Arrateis	approx. 1 pound/459 grams
Oytava	2 adarmes = ¼ ounce (6.36 grams)

Russian Measures

(Unless otherwise stated, from Clark, Daniel Kinnear, 'A Manual of Rules, Tables and Data for Mechanical Engineers', Blackie & Son, London, 1878, p. 172)

liniya (line)	= 0.1 inches (2.54 mm)
diuym	= 10 liniya (1 inch, 25.4 mm) (had been set precisely equal to an English inch by Peter the Great. See Cardarelli, F. 'Encyclopaedia of Scientific Units, Weights and Measures: Their SI Equivalences and Origins', 2nd ed., Springer, (2004), pp. 121–124. ISBN 1-85233-682-X)

Vershok	= 1.75 inches/44.45 mm.
Fuss (foot)	= 13.75 inches, 349.25 mm, (but after 1831, the English foot of 12 inches, 305 mm, was used)
Arshine /Arschine	= 16 vershoks = 28 inches, 711.2 mm (Gamel, 1826, p. 15)
Sagene/sajene	= 3 arschines = 7 feet/2.134 metres. (Gamel, 1826, p. 15)
Verst	= 500 sajenes = 3,500 feet, 1067.5 metres (Gamel, 1826, p. 2)
Zolotnik	= 65.833 grains, 0.1505 ounces, 4.25 grams
Pound	= 0.9 pounds avoirdupois, 14.45 ounces, 409 grams.
Pood	= 16.38 kg = 36 pounds avoirdupois approximately

The barrels of all Russian guns, in addition to those of the pistols, have a calibre of 7 lines for which lead bullets of six zolotniks in weight were used. (Gamel, translated edition 1988, p. 82, footnote)

Spanish Measures

Old units continued in use until 1859 when the French metric system was adopted. (Clark, 1878, p. 176, 177)

line (linea)	= 0.07725 inches (1.96 mm)
12 lineas	= 1 pulgada (0.927 inches, 23.55 mm)

Gun Calibre/Bore sizes.

Traditionally for smoothbore guns bore sizes have been designated as a number which represents the number of lead balls with diameters matching the bore and are equal to a weight of 1 pound. With the introduction of the rifle, size was generally expressed as the diameter of the bore in inches or millimetres.

The larger smoothbore sizes in this series are shown below, along with the Portuguese measure of weight, the adarme.

bore	**ball dia.**	**weight**	**weight**	
size	**inches**	**ounces**	**adarmes**	
5	.983	3.2	25.6	
6	.924	2.67	21.36	(20 adarmes = 2.5 oz)
7	.884	2.29	18.3	
8	.843	2.00	16	
9	.804	1.78	14.2	
10	.786	1.60	12.8	
11	.762	1.46	11.6	

bore	ball dia.	weight	weight	
size	inches	ounces	adarmes	
12	.747	1.33	10.64	
13	.734	1.33	10.64	
14	.713	1.14	9.15	
15	.703	1.07	8.56	
16	.661	1.00	8	
17	.654	0.94	7.5	
18	.649	0.89	7.1	
19	.643	0.84	6.7	
20	.630	0.81	6.5	

A bore of 14 adarmes means a ball of 1.75 ounces, equivalent to a bore size between English 9 and 10 bore (21.3 mm and 19.9 mm).

In the section dealing with methods described in *Espingarda Perfeyta* for finishing the bore, the adarme appears also to be used as a linear measure. There is a linear relationship between the adarme as a weight and the bore diameter. This is to be expected since ultimately bore size is based upon a weight. The relationship is best shown in graphical form.

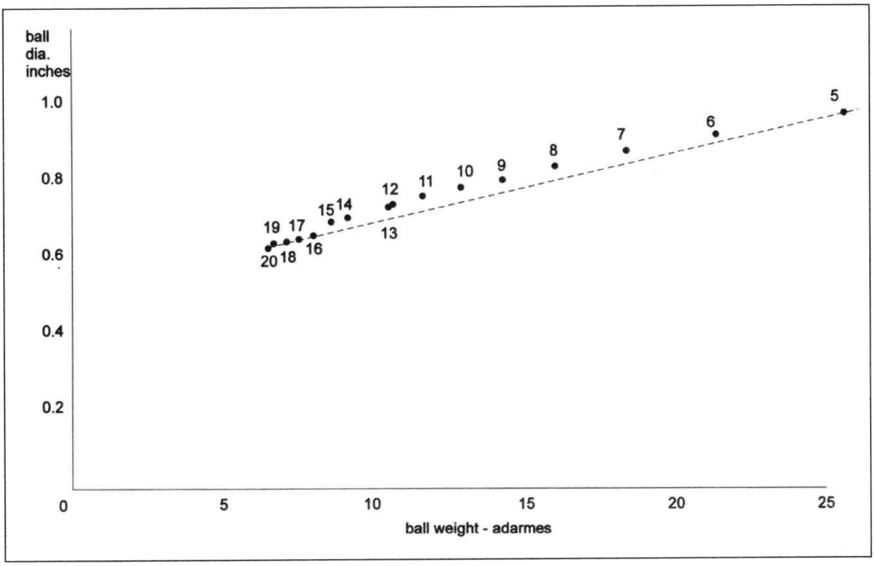

Graph showing the relationship between ball diameter in inches, bore size as numbers and weight in adarmes which can be seen as approximately linear.

Appendix 2

Mémoire important sur la fabrication des armes de guerre

[Important memoir on the manufacture of weapons of war]

H. Blanc

A l'Assemblée Nationale.

Printed L M Cellot, rue des Gr.-August, 1790

1. Il n'est rien sans doute qui intéresse aussi essentiellement l'Etat, relativement à la fabrication des armes de guerre, que la plus grande économie & la précision a plus exaĉte.
1. There is undoubtedly nothing that interests the State, in relation to the manufacture of weapons of war than the greatest economy & more exact precision.

2. C'est de cer objet important que le sieur Blanc n'a cessé de s'occuper depuis plus de trente annés sous les yeux du gouvernement, & c'est pour concourir, autant qu'il est en lui, a l'avantage public, qu'il va mettre au jour le résultat de ses travaux & de ses méditations.
2. It is with this important object that M. Blanc has not ceased to occupy himself for more than thirty years under the eyes of the government, & it is to contribute, as far as he is able, to the public advantage, that he is going to bring to light the result of his labours & his meditations.

3. Aprés avoir donné, pendant 15 années, toute son attention & son aĉtivité à l'étude de l'art de la fabrication des armes, le sieur Blanc sut nommé en 1763 à la place de côntroleur de la manufacture d'armes établie à Saint-Etienne en Forez.
3. After having given, for 15 years, all his attention & his activity to the study of the art of the manufacture of weapons, M. Blanc was appointed in 1763 to the position of controller of the manufacture of weapons established in Saint -Etienne in Forez.

4. L'imperfection où étoit alors ce genre de fabrication offert une vaste carriere à son zele & à ses soins pour la perfection de son art.
4. The imperfections at that time in this kind of manufacture offered a vast career for his zeal & his care for the perfection of his art.

5. Il s'occupa d'abord à reĉtifier & diriger le travail des platineurs, & quoique les ouvriers qu'il employoit sussent dispersés dans la campagne à une & deux ligues de l'éstablissement de la manufaĉture, il n'est pas moins parvenu à en faire de bons ouvriers.
5. He was at first occupied with rectifying & directing the work of the lock makers, & although the workmen whom he employed were dispersed in the country a league or two from the establishment of manufacture, he has nevertheless succeeded in making them good workers.

6. La trempe en paquet à été l'un des objets principaux de ses méditations & de ses expériences; & les lumieres qu'il a acquises sur ce genre de travail, que les Anglais ont porté a un si haut degré de perfeĉtion, l'ont mis en état d'opérer de maniere à ne rien craindre pour la divergence des pieces de platine, ni même pour celles de garniture, si l'on jugeoit comme il le croit, qu'il sût nécessaire qu'elles sussent trempées.
6. Case hardening has been one of the principal objects of his meditations & experiments; & the knowledge which he has acquired of this kind of work, which the English have carried to such a high degree of perfection, has enabled him to operate in such a manner as not to fear anything for the distortion of the lock components, nor even the furniture, if it were thought, as he believed, that it was necessary that they should be hardened or tempered.

7. Son application continuelle lui a fait découvrir les moyens 1. de perfectionner l'acier naturel qui nous vient de Rive, en le pêtrissant en trousse; & 2. de faire de l'acier de cémentation.
7. His continual application made him discover the means, firstly to perfect the natural steel which comes to us from Rive, by welding it in a bundle; & secondly to make cementation steel.

8. Aussi a-t-il eu l'avantage de mériter la confiance du gouvernement pour composer tous les modeles des fusils qui ont été faits pour les troupes en 1764, 1766, 1768, 1774 & 1777.
8. Also, he had the advantage of having the confidence of the government to furnish all the models of the muskets which were made for the troops in 1764, 1766, 1768, 1774, & 1777.

9. Il a inventé & fabriqué en 1777, un fusil destine pour le regiment du roi infanterie, sur l'invitation qui lui en a été faite par le lieutenant inspecteur; ce fusil sur présenté a M. de Gribeauval & au comié de la guerre pour y être

particuliérement examiné; & sur le rapport qui en a été fait an ministre par le comité, ce fusil obtint la préferentés sur tous les autres modeles qui avoient alors au government, & il sur adopté pour le service de toute l'infanterie de France.

9. He invented & manufactured in 1777, a musket destined for the royal regiment of infantry, at the invitation of the Lieutenant inspector; this musket, upon presentation to M. de Gribeauval, & to the committee of war, for its special consideration; & upon the report which was made to the minister by the committee, this musket became the one preferred over all the other models which the government had at that time & was adopted for the service of all the infantry of France.

10. Il s'est également occupé de la fabrication du pistolet, & il en a inventé un qui a beaucoup moins de pieces que ceux dont on avoit fait usage jusqu'alors, qui est infiniment plus solide, & qui exige beaucoup moins d'entrerien; le devant du canon est absolument sans bois ni garniture, & la piece de détente, le porte-vis, le corps de platine le bassiner & la bride, qui étoient précédemmeur en fer, & la partie du bois qui fert d'encastrement a toutes ces pieces, ne font, dans cette espèce de pistolet, qu'un corps en cuivre: qui recoit de la fonte la forme nécessaire.

10. He has equally occupied himself with the manufacture of the pistol, & has invented one which has fewer pieces than those which had hitherto been used, which is infinitely more solid, & which requires much less maintenance; the front of the barrel is absolutely without wood or trim & is replaced by a brass casting, made in an iron mould, with integral lockplate & pan with its bridle, formerly made of iron, & to which are attached the trigger & other lock components.

11. L'utilité des découvertes qu'avoit faites le sieur Blanc parut tellement importante au comité, que le gouvernement ne crui pouvoir mieux faire que de créer, pour lui procurer les moyens d'être encore plus utile a la fabrication des armes, la place de contrôleur principal des trois manufactures du royaume, qu'il exerce depuis le 18 mars 1778.

11. The usefulness of the discoveries made by M. Blanc appeared so important to the Committee that the Government could not do better, in order to implement them, than to appoint him, so to be even more useful to the manufacture of arms, the chief inspector of the three manufactories of the kingdom, which was done on 18th March 1778.

[intimation of use of gauges]

12. La premiere operation du sieur Blanc, en cette nouvelle qualité, sut de faire un réglement des proportions des pieces & mesures de l'arme , & des mesures & calibres pourr servir de vérificateur aux contróleurs subalterns; &

le régiement, les mesures, & les modeles servirent de bases au travail de ces trois manufactures.

12. The first operation Mr Blanc, in this new capacity, was able to regulate was the proportions of the pieces & measurements of the weapon, & measurements and calibres to serve as verifier to the assistant controllers; & the regulations, the measures, & the models served as a basis for the work of these three manufactories.

13. Le sieur Blanc s'occupa ensuite de l'invention de differentes mesures, outils, machines & matrices inconnus jusqu'alors dans nos manufactures en nombre assez considerable pour faciliter le travail de plus de 800 ouvriers & donner a ce travail toute la precision dont il étoit susceptible.

13. M. Blanc then occupied himself with the invention of various measures, tools, machines, & dies, hitherto unknown in our manufactures, in considerable number to facilitate the work of more than 800 workers, & to give to this work all the precision possible.

14. Il visita en 1782 les manufactures de Charleville, de Maubeuge & de Tulle; il sat frappé de la multiplicité d'imperfections que la negligence avoit laisse-subsister dans la fabrication des armes, & il se livra a toutes les reflections capables de lui suggérer les moyens d'y remédier.

14. He visited the manufactories of Charleville, Maubeuge, & Tulle in 1782; he was struck by the multiplicity of imperfections which negligence had left in the manufacture of arms, & he gave himself up to every reflection capable of suggesting to him the means of remedying it.

15. C'est ainsi que le sieur Blanc est parvenu a créer un plan de travail qui embrasse toutes les pieces de l'arme, & assure tellement l'opération de l'ouvrier, par le moyen des machines & outils qu'il a fait fabriquer, que le plus mauvais ouvrier se trouve dans l'impossibilité de mal faire.

15. It is thus that M. Blanc has succeeded in creating a plan of work which embraces all the parts of the weapon, & so ensures the operation of the workman by means of the machines and tools he has made, with which it is impossible for the worst worker to make errors.

16. A cette époque , la fabrication des canons de fusils étoit tellement vicieuse, qu'il s'en trouvoir toujours au moins un quart de rebutés en fortant des mains de l'ouvrier, en forte que l'on étoit tous les jours a la veille de manquer de canons pour completter les d'armes nécessaires au service de l'armée.

16. At that time, the manufacture of gun barrels was so vicious that there were always at least a quarter of them rejected by relying on the hands of the worker, as we were every day on the point of running out of barrels to complete the weapons necessary for the service of the army.

17. C'est alors qu'il imagina une nouvelle maniere de forger les canons de fusils, dont le résultat a été de donner au fer plus de densité, & de le render plus homogene en le foudant plus également & plus intimement, indépendamment de ce que ce nouveau procédé rend les parois du canon d'une égalité qu'il est physiquement impossible d'obtenir par les methods qui étoient usitées précédemment, & qui ont entr'autres inconvéniens celui de laisser multiplier dans les canons des cavités que l'on appelle chambres.
17. It was then that he devised a new way of forging gun barrels, the result of which was to give the iron more density, & to make it more homogeneous by welding it more evenly & more intimately, & this new process gives the walls of the barrel an equality which it is physically impossible to obtain by the methods which were used previously, & which have among other disadvantages that of allowing cavities to multiply in the barrels.

18. Par ce nouveau procédé, le sieur Blanc a eu d'excellents canons, quoique forges d'ailleurs avec du mauvais fer, tandis qu'avec de l'excellent fer il ne laisse pas que d'y avoir beaucoup de mauvais canons dans les manufactures où l'on suit encore les anciens procédés.
18. By this new process, M. Blanc has made excellent barrels, even when forged, with poor iron; while even with excellent iron many poor barrels are made using the former processes.

19. Aussi le sieur Blanc a-t-il eu l'avantage que sur la forge des canons selon sa méthode, il s'en trouvoir a peine neuf a dix sur cent dans le cas du rebut, tandis que dans les autres manufactures où l'ancienne fabrication est toujours suivie, elles continuent a éprouver une perte constante de 25 a 30 pour cent fur les canons fabriqués.
19. Hence M. Blanc has the advantage that on the forging of barrels according to his method, there is scarcely nine to ten per cent rejected, while in other manufactories, where the former method is used, they continue to experience a constant loss of 25 to 30 per cent of the barrels made.

20. Il est facile de juger par ce premier apercu de la diminution de dépense dans la fourniture de fusils aux troupes, qui résulte évidemment du procédé du sieur Blanc pour la fabrication des canons de ces fusils.
20. It is easy to judge by this first glimpse of the diminution of expense in the supply of muskets to the troops, which evidently results from the process of M. Blanc for the manufacture of the barrels of these muskets.

21. Aussi M. Duchatelet, qui avoit, ainsi que M. de Gribeauval, pris la peine de se convaincre de l'avantage & de l'importance de la méthode du sieur Blanc, s'empresserent-ils de render compte au ministre des succés de cet artiste; & le ministre, pour faciliter les effets du zele & de l'ntelligence du sieur Blanc,

lui accorda un logement au donjon de Vincennes ou il a formé un nouvel établissement depuis le mois de mai 1786.

21. Also, M. Duchatelet, who, like M. de Gribeauval, took the trouble to convince himself of the advantage & importance of the method of M. Blanc, hastened to report to the minister of his successes; & the minister, to facilitate applying the results of the zeal & intelligence of M. Blanc, granted him lodging at Vincennes, where he had formed a new establishment in the month of May 1786.

[interchangeable locks]

22. Il s'y est dés-lors occupé principalement de la fabrication des platines, & a inventé pour cela plusieurs machines & un grand nombre d'outils don't l'utilité est telle, que les pieces qui composent une platine étant fabriquées separément, peuvent, (après qu'elles auront été achevées) être mêlées & confondues ensemble, & ensuite être prises chacune au hazard, pour servir a former la platine, au moyen de quoi il n'est personne qui ne soit en état de remonter une platine avec la même facilité que s'il ne s'agissoit que de changer la pierre.

22. He has since occupied himself chiefly with the manufacture of locks, & has invented for this purpose several machines & a large number of tools of such usefulness that the pieces composing a lock being manufactured separately, may, after they have been completed, be intermingled & confused together, & then each taken at random, to assemble the lock, by means of which there is no one who is not able to reassemble a lock, with the same ease as if it were only a question of changing the flint.

23. Le sieur Blanc pourroit a cet égard invoquer le témoignage de M. de Brienne & celui des officiers généraux qui l'accompagnerent a Vincennes pour s'assurer des succes des travaux du sieur Blanc puisqu'ayant pris au hasard dans son atelier les différentes pieces nécessaires pour fourmer une platine, ce ministre en monta une sans aucune difficulté.

23. In this respect M. Blanc may invoke the testimony of M. De Brienne, & that of the general officers who accompanied him to Vincennes to assure himself of the success of the work of M. Blanc, since, having taken at random in his workshop the various pieces necessary to create a lock, the minister assembled one without any difficulty .

24. L'on concoit de-la l'avantage inappreciable qui résulte de la méthode du sieur Blanc, car d'une part-il n'est point a craindre que lorsqu'une piece vient a être cassé dans une platine, l'ouvrier qui est chargé de la raccommoder soit exposé a l'inconvénient d'altérer aucune des autres pieces qui la composent, en voulant rétablir celle qui a été cassée, & adapter la nouvelle a la platine, puisque.

24. We conceive of the invaluable advantage which results from the method of M. Blanc, for, on the one hand, it is not to be feared that when a piece comes to be broken in a lock, the worker who is obliged to repair it, has the inconvenience of altering any of the other pieces which compose it, by replacing the one which has been broken, & to adapt the new to the lock.

25. Toutes celles qui résultent de l'invention du sieur Blanc peuvent servir sur le champ & sans choix a remplacer celle qui ne peut plus servir, & de l'autre, unsoldat ou toute autre personne qui connoitra seulement les pieces de la platine, sera en état de remplacer la piece manquante par la premiere qui lui sera fournie par le garde-magasin de l'armée, en forte que l'on est dispensée par-la de toute espece d'emménagement, soir dans les armées qui font en marche, soit sur les vaisseaux de guerre pour tout ce qui regarde la réparation des armes.

25. All those which result from the invention of M. Blanc may be used on the field, & without special selection replace that which is broken, & a soldier or any other person who knows only the pieces of the lock, can replace the broken piece by that supplied at random by the store-keeper of the army in the field or on the ships of war all that relates to the repair of arms.

26. C'est pour constater ces avantages que le sieur Blanc a d'abord fabriqué les pieces nécessaires a former cent platines, qu'il les a mises sous les yeux de M. de Gribeauval avant qu'elles sussens trempées, qu'il leur a ensuite donné la trempe, les a polies, les a représentées, & a manifesté d'une maniere évidente que l'assemblage des pieces n'étoit pas moins facile après la trempe qu'auparavant.

26. In order to ascertain these advantages, M. Blanc first manufactured the pieces necessary to form a hundred locks, which he put before the eyes of M. de Gribeauval before they were hardened, then hardened & tempered them, polished them, re-presented them, & manifested in an evident manner, that the assembly of the pieces was not less easy after the hardening & tempering than before.

27. Mais comme ce n'est qu'en réitérant les expériences que l'on peut obtenir une pleine conviction sur les résultats d'un art méchanique, & acquérir la certitude de ses succés, & qu'il est d'ailleurs difficile, pour ne pas dire impossible, de s'assurer de la durée d'un outil ou d'une machine definés a faite un service journalier, & dont les effets doivent avoir toujours la même precisión, lorsque l'on n'en a employé l'usage que pendant quelque temps, le sieur Blanc, jaloux de prouver l'efficacité & la solidité de ses moyens, ne s'est pas borné a la fabrication des différentes pieces qui peuvent servir, sans aucun choix, a monter cent platines, il a enrepris, a 'laide des ouvriers qu'il a formés a Vincennes, de faire une affez grande quantité de pieces pour en monter mille; & ce travail que le sieur Blanc a entrepris sous l'autorite du gouvernement

est achevé. Cinq cents de ces platines sont déja déposées chez le ministre a l'hôtel de la guerre, & les cinq cents autres sont encore dans les atteliers du sieur Blanc au donjon de Vincennes.

27. But as it is only by repeating the experiments that one can obtain full conviction on the results of a mechanical art, & acquire the certainty of its successes, & that it is besides, difficult, not to say impossible, to ascertain the duration of a tool or machine in daily operation & which must retain its accuracy. Over some time, M. Blanc, anxious to prove the efficacy & solidity of his method, has not confined himself to the manufacture of the various parts which can serve, without any choice, to accommodate a hundred locks; he has, with the help of the workers he has trained at Vincennes, to make pieces for a thousand; & this work which M. Blanc undertook under the authority of the Government is completed. Five hundred of these locks are already deposited with the Minister at the Hotel de la Guerre, & the other five hundred are still in the quarters of M. Blanc at Vincennes.

28. Le sieur Blanc peur même ajourer qu'il a personnellenent acquis la conviĉtion que la dépenss du travail des platines sera beaucoup moindre que celle que ce travail occasionne dans les manufaĉtures, & que la dépense des réparations sera réduite a la moité.

28. M. Blanc is afraid to admit that he has personally acquired the conviction that the expense of the work of the locks will be much less than the expense of this work as currently carried out in the manufactories, & that the expense of the repairs will be reduced to one half.

29. Il faut maintenant-examiner si la position & l'état des des manufactures qui existent actuellement en France font susceptibles d'y recevoir le travail du sieur Blanc.

29. We must now consider whether the position & condition of the manufactories which now exist in France are likely to adopt the work of M. Blanc.

30. Quatre manufactures d'armes a feu, & une manufacture d'armes blanches, forment toutes nos ressources pour le service de nos armées de terre & de mer.

30. Four manufactories of fire-arms, & a manufactory of edged weapons, form all our resources for the service of our armies of land & sea.

31. Ces manufactures font établies a Saint-Etienne-en-Forez, a Charleville en Champagne, a Maubeuge en Hainault, a Tulle en Limousin, & la manufacture d'armes blanches est a Cligental en Alsace.

31. These manufactories were established at Saint-Etienne-en-Forez, at Charleville in Champagne, at Maubeuge in Hainault, at Tulle in the Limousin, & the manufactory of edged weapons at Klingenthal in Alsace.

32. Il s'en faut bien que les situations des manufactures de Charleville & de Maubeuge soient favourable a la fabrication. Il n'y a pas de carrieres de meules dans le pays où ces manufactures font situées, & l'on est obligé dentirer a grand prix du pays de Liege.
32. The situation of the manufactories of Charleville & Maubeuge is far from being favourable to manufacture. There are no quarries for grindstones in the country where these manufactories are situated, & one is obliged to pay a great price for them from the country of Liège.

33. Elles manquent également l'une & l'autre de bois de noyer, qui est indispensable pour l'établissement de l'arme a feu, & il faut, pour s'en procurer, venir jusques aux portes de Paris.
33. They are equally lacking in walnut wood, which is indispensable for the manufacture of fire-arms, & it is necessary, to procure it, to come as far as the gates of Paris.

34. Il existe dans ces manufactures un autre inconvénient, celui que les usines qui fervent a leur exploitation, n'ont pu être construites qu'a une & deux lieues de l'établissement des manufactures par rapport au local.
34. There is another disadvantage in these manufactures, that the factories which would adopt these methods cannot be constructed except one or two leagues from the existing manufactories.

35. Enfin ces deux établissements, étant loin du centre du royaume, & prés des frontieres, ne font peuplés que d'ouvriers liégeois qui, conservant toujours l'esprit de retour dans leur patrie ne peuvent être solidement attachés a ces manufactures.
35. Finally, these two establishments, being far from the centre of the kingdom, & near the frontiers, are populated only by Liège workers, who, always retaining the desire to return to their country, cannot be firmly relied upon to remain at these manufactures.

36. La manufacture établie a Tulle, & dont les travaux font spécialement destinés au service de la marine, a ses usines dans un lieu trés-mal fain, & les ouvriers ont beaucoup de peine a s'y habituer, au moyen de quoi cetre manufacture fournit a peine six mille armes par an.
36. The manufacture established at Tulle, the work of which is specially designed for the service of the navy, has its factories in a very bad place, & the workmen have great difficulty in accustoming themselves to it, resulting in this manufactory furnishing hardly six thousand weapons per year.

37. Le pays abonde, a la vérité, en noyers & a d'excellent fer; mais l'on n'y a découvert du charbon de terre que depuis quelques années, & il est d'une trés-médiocre qualité.

37. The country abounds, indeed, in walnut & excellent iron; but coal has been discovered for only a few years & is of very poor quality.
38. Cette manufacture est d'ailleurs éloignée des rivieres navigables, & il n'y a conséquemment pas de facilité pour le transport des armes & des matieres nécessaires pour la fabrication.
38. This manufactory is, moreover, far from navigable rivers, & consequently there is no facility for the transport of arms & the materials necessary for their manufacture.

39. La manufacture établie a Saint-Etienne présente, a la vérité , plus d'avantages, parce qu'elle est environnée de montagnes qui renferment une grande quantité d'excellent charbon fossille.
39. The manufactory established at Saint-Etienne presents, in truth, more advantages, because it is surrounded by mountains which contain a great quantity of excellent fossil coal.

40. Elle est d'ailleurs voisine du Rhône & de la Loire qui facilitent le transport de ses productions dans le Nord & dans le Midi , & elle est a la proximité du Dauphiné & du Vivarais, où les noyers font abondans.
40. It is nevertheless near the Rhone & the Loire, which facilitates the transport of its products in the North & in the South, & is near the Dauphine & Vivarais, where the walnut is abundant.

41. Mais ces avantages mêmes ont donné lieu a l'établissement d'une autre manufacture d'armes, uniquement destinée pour le commerce , & la proximité de ces deux manufactures occasionne entr'elles tous les inconvéniens de la rivalité, qui nuisent continuellemement a la activité de la fabrication des armes de guerre.
41. But these very advantages have given rise to the establishment of another manufactory of arms, which is destined only for commerce, & the proximity of these two manufactories occasioned all the inconveniences of rivalry, which continually hindered the activity of the manufacture of weapons of war.

42. Quant a la manufacture d'armes blanches établies a Clingental, son rapport avec les manufactures d'armes a feu consiste en ce qu'elle fournit a celles-ci des bayonnettes.
42. As for the manufacture of edged weapons established at Klingenthal, its relation with manufactories of firearms is that it furnishes them with bayonets.

43. Mais se trouvant placée a prés de cent lieues des quatre Manufactures d'armes a feu, elle ne peut leur fournir les bayonnettes qu'avec des frais considérables de transport.
43. But being placed nearly a hundred leagues from the four manufactories of firearms, it can furnish them bayonets only with considerable expense of transport.

44. L'on concoit enfin que cette manufacture étant, ainsi que celles de Charleville & de Maubeuge, prés des frontieres, elle n'est pas moins exposée que celles-ci aux incursions de nos voisins, & conséquemment aux événemens désastreux qui pourroient être les suites d'une guerre.
44. It is conceived, moreover, that this manufactory being, as well as those of Charleville & Maubeuge, near the frontiers, is not less exposed than these to the incursions of our neighbours, & consequently to the disastrous events which might be the consequences of a war.

45. L'on ne peut se dissimuler encore que le nombre des manufactures, & la distance qui se trouve enntr'elles, en rend la surveillance difficile & tres-coûteuse.
45. It cannot be concealed, however, that the number of manufactories, & their distance, renders the surveillance difficult & very costly.

46. Il est sensible d'ailleurs que les principes de fabrication ne peuvent être uniformes dans les manufactures don't le plus grand nombre des ouvriers sont étrangers, & ont, été imbus de mauvais principes qu'ils abandonnent avec peine, & que leur indocilité naturelle dispose a chaque instant a se livrer au plus offrant.
46. It is, moreover, sensible that the principles of manufacture cannot be uniform in the manufactories of which the greatest number of workmen are foreign, & have been imbued with bad principles, which they abandon with difficulty, & that their natural indocility has always led them to work for the highest bidder.

47. Un vice encore plus suneste dans nos manufactures d'armes, c'est qu'elles ont été livrées a l'entreprise de gens qui se font bien moins occupés du bien du service que des moyens qui pouvoient augmenter leur fortune.
47. An even more vile defect in our manufactories of arms is that they have been handed over to the undertaking of people who are far less concerned with the good of the service than with the means of increasing their wealth.

48. Ainsi, tandis que des traitans sans principes, comme sans connonoissance de la fabrication des armes, uniquement avides de gain, saisoient en peu de temps, dans l'entreprise des manufactures d'armes de guerre, des fortunes immenses, l'art de l'arquebuserie est resté dans ces manufactures en un état d'engourdissement qui ne permettoit aucun des progress que le Sr Blanc leur a fait faire, lorsque sa réputation l'a eu place au-deffusdes dégoûts & des diffcultés sans nombre que lui ont prodigués ceux qui avoient intéret a laisser subsister l'ancien régime des manufactures d'armes.
48. Thus, while traitors without principles, & without knowledge of the manufacture of arms, only avid for gain, seized, in a short time, through the enterprise of manufactories of weapons of war, immense fortunes, the art of gunmaking has remained stagnant in these manufactories which did not

permit any of the developments which M. Blanc has made to be introduced, because of the numberless difficulties placed in the path of those who had an interest in leaving behind the old regime of arms manufacture.

49. Mais, en supposant que ces entrepreneurs eussent quelque théorie, cet avantage seroit insuffisant, s'il n'etoit joint a celui qui résulte de la pratique, parce qu'il n'y a que la pratique qui puisse, dans les arts méchaniques, éclairer les operations & en assurer la perfection.

49. But, supposing these contractors to have some theory, this advantage would be insufficient if it were not joined to that which results from practice, because there is only practice which can, in the mechanical arts, understand the processes & ensure their perfection.

50. C'est par cette raison que le meilleur ouvrier est toujours le plus adroit, tandis que l'homme qui n'a que la théorie peut a chaque instant donner dans l'erreur, en croyant tendre a la perfection, parce qu'il ne connoit pas l'effet de la main-d'oeuvre.

50. It is for this reason that the best worker is always the most skilful, while the man who has only theory can at any moment create error when he believes he tends to perfection, because he does not know the value of practical skill.

51. Il est évident, d'aprés cela, que si celui qui est a la téte d'une manufacture d'armes, comme de toute autre ne réunit pas les connoissances de la pratique aux lumieres de la théorie, il est impossible que rien de parfait forte de la manufacture dont il a entrepris l'exploitation, parce qu'il est incapable de juger & de rectifier les opérations du mauvais ouvrier, & qu'il l'est également de concevoir & de s'approprier les idées qui lui font présentées par l'artise éclairé qui joint a la pratique l'intelligence néceffaire pour se rendre compte de ses opérations.

51. It is evident from this that if he who is at the head of an arms factory, like any other, does not unite the knowledge of the practice to the light of theory, it is impossible anything will be perfect for he is incapable of judging & rectifying the operations of the bad workman, & is equally incapable of conceiving & appropriating the ideas which are presented to him by the enlightened artisan who joins to practice the necessary awareness of inefficiency in its operations.

52. Chacun des ouvriers qui sont employés dans les manufactures ne connoit pas le travail de toutes les parties qui entrent dans la composition de l'arme, parce que le travail de ces différentes parties est divisé entr'eux, les uns n'étant occupés qu'a la fabrication des canons de fusils, les autres a la preparation des bois, a la sacon des garnitures, & enfin au travail des platinese, en forte que le chef de la manufacture se trouve placé au milieu de gens qui ne peuvent respectivement lui fournir que des idées incompletes, d'ou il résulte que

des manufactures ainsi organisées ne trouvent aucun moyen de concourir a la perfection des armes, & qu'il y a nécessairement lieu a de frequentes réparations, sut-tout pour les platines dont deux pieces essentielles, les vis & les ressorts, sont trés-casuelles.

52. Each of the workmen who are employed in manufactories does not know the work of all the others who make the parts which comprise the weapon, because the work of these different parts is divided among them, The manufacture of barrels for guns, others for the preparation of wood, the making of furniture, & finally the work of lock makers, so that the chief of the manufactory is placed in the midst of people who cannot supply him more than incomplete ideas, from which it follows that manufactories thus organised find no means to contribute to the perfection of arms, & that there is necessarily occasion for frequent repairs, especially for the locks, of which two essential parts, screws & springs, are very common.

53. La réparation des vis feroit moins difficile, s'il n'étoit notoire qu'il n'existe pas dans les atteliers des arquebusiers deux filieres, dont les tarauds soient exactement de la méme grosseur, & l'espacement des filets parfaitement égal.

53. The repair of the screws would be less difficult if it were not that there are not two screws in gunmakers' work having threads which are exactly the same size, & the spacing of the threads perfectly equal.

54. Au moyen de cet inconvénient, il n'y a pas d'autre ressource pour un arquebusier qui doit remplacer une vis manquante a une platine, que de se servir pour filiere du trou de la platine auquel manque la vis, pour tarauder la nouvelle vis qui doit la remplacer.

54. By means of this disadvantage, there is no other resource for a gunmaker that must replace a missing screw in a lock, to use the hole in the lockplate as a die to thread the new screw.

55. Mais si le corps de la platine n'a pas été bien trempé, & quelle morceau de fer destine a faire une nouvelle vis ne soit pas bien recuit, l'on risque que le filet du corps de platine soit altéré, s'il n'est entiérement détruit par l'effort que l'on est obligé de faire pour former le filet de la nouvelle vis, & en ce cas, qui n'est que trop ordinaire, il n'y a pas d'autre moyen que de recuire entiérement le corps de platine pour tarauder avec un plus gros taraud celui des trous qui a été ainsi endommagé, & cette nouvelle opération produit de nouveaux inconvéniens:

55. But if the body of the plate has not been well tempered, & the piece of iron intended to make a new screw is not well annealed, there is a risk that the thread in the lockplate may be entirely destroyed by the effort which is necessary to form the thread of the new screw, & in this case, which is only too common, there is no other means than that of using a larger tap in that hole which has been so damaged, & this new operation creates new drawbacks:

A. La perfection de la platine est altérée par l'agrandissement du trou destiné a recevoir la nouvelle vis.
A. The perfection of the plate is altered by the enlargement of the hole to receive the new screw.
B. La scorification que détermine le recuit, sur-tout dans les trous des pivots & des vis qui sont pratiqués dans le corps de platine, occasionne le balottement des pieces & opere dans peu de temps la destruction totale de la platine.
B. The scorification which is caused by annealing, especially in the holes of the pivots & screws which are made in the lockplate, causes the parts to be loose, & in a short time the complete destruction of the lock.
C. Enfin, le prix de la réparation se trouve doublé par l'indispensable nécessité de tremper la nouvelle vis, le corps de platine & la piece a laquelle il faut l'adapter.
C. Finally, the price of the repair is doubled by the indispensable necessity of hardening & tempering the new screw, the lockplate & the piece to which it must be adapted.

56. Les mêmes inconvéniens se reproduisent, lorsqu'il est question de refaire la noix, la gachette, la bride, ou même le chien, qui a cassé dans une platine, parce qu'il est physiquement impossible a un arquebusier, quelqu'habile qu'il soit, de réussir, dans cette opération, a accorder ensemble toutes les pieces sans toucher a celles qui avoisinent la piece a remplacer.
56. The same inconveniences are repeated when it is a question of replacing the tumbler, the trigger, the bridle, or even the cock, which has broken in a lock, because it is physically impossible for a gunmaker, in this operation, to succeed in putting all the parts together without altering those which adjoin the piece to be replaced.

57. Il n'y a pas moins de dificulté dans les réparations du ressort, parce qu'il faut pour cela avoir de la matiere premiere de la meilleure qualité, & beaucoup d'adresse & d'intelligence pour le forger, pour le tremper, sur-tout pour le recuire, & eu supposant toutes ces opérations bien faites, aucun ouvrier ne peut encore répondre de la durée d'un ressort.
57. There is no less difficulty in the repairing of the spring, because it requires, for that purpose, the raw material of the best quality, & much skill and intelligence to forge it, harden & temper it, &, supposing all these operations well done, no worker can yet answer for the duration of a spring.

58. L'on concoit facilement dela a quel embarras les troupes peuvent se trouver exposées pour la reparation de leurs armes, losqu'elles feront éloignées des manufactures, dautant qu'il y a à peine un arquebusier sur vingt, de oeux qui font établis dans les différentes villes du royaume, qui soit en état de faire

complettement une platine; & ceux qui la savent faire font trop occupés a la monture & aux pieces de garniture pour pouvoir acquérir assez de routine pour bien faire les reparations des platines & les entreprendre au prix modéré que l'on paye aux manufactures. Il n'est personne sans doute qui ne soit en état de juger combine cet inconvenient est préjudiciable aux réparations des armes, sur-tout de celles de la marine, don't le besoin pressant & instantané oblige le gouvernement d'user d'une espece de violence pour faire venir quell-quesois de fort loin des ouvriers qui ne font dans le fait que des apprentiss qui parcourent les provinces pour se perfectionner dans leur métier.

58. It is easy to conceive of what embarrassment the troops may be exposed for the repair of their arms, when they are far removed from manufactories, inasmuch as there is scarcely one gunmaker out of twenty cities of the kingdom, who is able to make a complete lock; & those who know how to do it are too busy making the other parts to be able to acquire skills in order to repair the locks & to undertake them at the moderate price that is paid at the manufactories. No one, who is in a condition to judge, is in any doubt how unfavourable this inconvenience is to the repairs of the pieces, especially those of the navy, whose urgent and instantaneous necessity obliges the government to conscript workers from great distances & who are in fact only apprentices who traverse the provinces to perfect themselves in their trade.

59. Il n'est d'ailleurs aucun art où il faille une si grande quantité d'outils que celui de l'arquebuserie, dont la perfection de l'euvrage est toujours en raison du nombre & de la perfection des outils, qui ne peuvent être bien faits que par les arquebusiers eux-mémes, ce qui exige de leur part beaucoup plus d'adresse que pour faire complettement une platine, operation qui nécessite des emménagement stables & solides, pour placer une forge & les gros outils nécessaires pour la fabrication de ceux qui ont moins de volume, & ce qui est impracticable a l'armee & particuliérement sur les vaisseaux.

59. There is, moreover, no art in which there is so great a need for tools as that of gunmaking, the perfection of which is always due to the number & perfection of the tools which cannot be better made than by the gunmakers themselves, which requires on their part much more skill than to make a complete lock, an operation which requires a stable & solid base to place a forge, & the great tools, which is impracticable to the army, & especially to the navy, but necessary for the manufacture of the smaller tools & parts.

60. Il est bon d'observer que l'usage ordinaire des platineurs est de donner en même-temps la forme a toutes les pieces destinées a la composition d'une platine, & que par consequent pour faire accorder ces pieces, l'on est dans l'usage de la ffer du fer a une piece, pour remplacer celui qui manque a une autre piece, & que fouvent l'ouvrier a mal-adroitement ôté de trop, ce qui prove l'impossibilité de trouver parmi quelques platines deux pieces égales que l'on puiss changer au besoin.

60. It is well to observe that the ordinary construction of locks is to give at the same time the form to all the parts intended for the composition of a lock, & that consequently, in order to make these pieces alike, the use of the part in one piece, in order to replace that which is missing in another piece, & which the workman has often improperly removed, which leads to the impossibility of finding among two locks two pieces which can be changed as needed.

61. Le travail inventé par le sieur Blanc a au contraire pour objete essential de faire des pieces détachées propres a former des platines, & auxquelles on donne la forme definitive avec des machines ou des matrices dont la régularité ne peut varier, ce que l'on ne peut attendre de la main seule des ouvriers; & par ce moyen l'on jouirit de l'a l'avantage inappreciable de porter dans une caisse des pieces de rechange de toutes especes, trempées, polies, & prétes a remplacer la piece qui manqueroit, enforte qu'il ne feroit plus nécessaire que d'avoir une personne chargée de nettoyer les armes, pourvu qu'elle sùt bien les démonter, ce que l'on peut facilement trouver par-tout.

61. The work invented by M. Blanc has, on the contrary, the essential object of making separate pieces suitable for forming locks, & to which we give the definitive form with machines or dies, the regularity of which cannot vary, & which cannot be expected from the hand alone of the workmen; & by this means one would enjoy the inestimable advantage of carrying in a crate spare parts of all kinds, hardened & tempered, polished, & ready to replace the piece which would be wanting, since it would no longer be necessary to have a person in charge of cleaning the arms, provided that they can dismantle them, which can easily be taught.

62. Le sieur Blanc a également établi le travail pour faire les garnitures d'aprés le mode de fabrication pour les platines, de maniere que ces pieces, qui sont comme celles de la platine, sujettes a se casser, pourrent être aussi promptement & aussi facilement remplacées que celles qui forment la platine, & l'on aura de plus, par cette méthode, l'avantage d'assurer l'uniformité du travail de toutes les parties du canon, & sur-tout celles des douilles de bayonnettes, avec les bouches de ces mêmes canons, ce qui épargnera les frais immenses de transport, que l'on est obligé de faire pour les bayonnettes, attendu qu'elles ne se fabriquent pas dans les manufactures d'armes a feu.

62. M. Blanc has also established the work to make the fittings according to the manufacturing method for the locks, in the same way that these parts, which are like those of the lock, subject to breakage, can also be promptly & easily replaced & that, by this method, we shall have the advantage of assuring the uniformity of the work of all the parts of the barrel, & especially those of the bayonet sockets & muzzles of these same barrels, which will spare the immense expense of transport, which we are obliged to make for bayonets since they are not manufactured in manufactories of firearms.

Rapport Fait a l' Academie Royale des Sciences

[Report made at the Royal Academy of Sciences]

D'un Mémoire Important, De M. Blanc, sur la fabrication des Armes de guerre.

[Ref: M. Blanc, *An Important Memoir on the Manufacture of Weapons of War.*]

le Roy, Laplace, Coulomb, Borda

Le Samedi 19 Mars 1791.
Extrait des Registres de l'Academie Royale des Sciences.
Du 19 Mars 1791

1. M. Blanc, Contrôleur principal des trois Manufactures d'Armes au service de l'artillerie, a présenté a M. de la Tour-du-Pin, Ministre de la Guerre, une *Memoire important sur la fabrication des Armes de guerre.* Ce Ministre en a renvoyé l'examen a l'Academie, qui nous a chargé, MM. de Borda, de la Place, Coulomb & moi, de lui en render compte. C'est ce que nous allons faire en entrant dans les details propres a metre la Compagnie a portée de juger, du projet de l'Auteur, sur la fabrication des armes a feu, des moyens qu'il propose pour le mettre en execution, & des avantages qui peuvent en résulter.
1. M. Blanc, Chief Controller of the Three Manufactories of Arms in the Service of Artillery, presented M. de la Tour-du-Pin, Minister of War, with an important memoir on the manufacture of weapons of war. This Minister has referred the matter to the Academy, who has instructed us, MM. De Borda, de la Place, Coulomb, & myself [le Roy], to give him an account of it. This is what we are going to do by entering into the details proper to the Company to judge the author's plan on the manufacture of firearms, the means he proposes to put it into execution, & advantages that may result.

2. Les armes a feu sont sujettes, comme tout le monde le sait, a un grand nombre d'accidens; mais les plus fréquens viennent de leurs platines, don't tantôt une pièce, tantôt une autre, s'usr ou fe casse, & par-la en fait manquer les effets, ou met ces armes hors de service.
2. Firearms are subject, as everybody knows, to a great number of accidents; but the most frequent of them come from their locks, & sometimes one piece, sometimes another, is destroyed or broken, so it does not function correctly & puts these arms out of service.

3. Quelquesois aussi on ne peut en faire usage, parce que quelques-unes de ces pieces sont égarées ou perdues.
3. Some also cannot be used because some of these pieces are lost.

4. Or, c'est particulièrment dans les armes des troupes qu'on voit ces divers accidens se renouveler le plus souvent, soit par la fatigue, si cela se peut dire, qu'éprouvent certaines pièces des platines dans les fréquens exercices, soit par la negligence du Soldat, peu habile en générale a manieres fortes de machines: de la nombre d'armes deviennent inutiles, qui pourroient encre être employees avec avantage, en refaisant ou en y remettant la pièce, ou les pieces qui manquent.
4. Now it is particularly in the arms of the troops that we see these various accidents repeated most often, either by the fatigue, if it may be so, which certain parts of the locks experience in the frequent exercises or the negligence of the soldier, who is generally not very clever in the way of machines, so a number of weapons become useless, which could be employed with advantage, by repairing or restoring the piece, with the parts which are needed.

5. C'est ce qu'on a si bien reconnu dans l'armée, qu'il y a plusiers Regimens qui, depuis un certain nombre d'années, ont soin de fe munir de pieces de rechange, afin de pouvoir, dans l'occasion, les substituer, dans les platines, a celles qui exigents d'être remplacées; mais il est facile de voir combine ces pieces, faite sans méthode, doivent mal se raccorder avec celles des platines auxquelles on veut les adapter, & les peines qu'il faut prendre pout les y adjuster de maniere, qu'elles remplissent exactement le même objet que celles auxquelles on les substitue.
5. This is what has been so well recognized in the army, that there are several Regiments, who for a number of years have been careful to furnish themselves with spare parts in order to be able to replace in the locks those parts which require replacement; but it is easy to see that these parts, made without method, are combined with those of the locks to which they have to be adapted, & have to be adjusted in such a manner that they work in exactly the same manner as those for which they are substituted.

6. On fentira facilement, d'après ces diverses observations, sans que nous soyons oblige d'y en ajouter d'autres, de quelle importance il feroit pour l'armée, qu'on pût fabriquer les platines de fusils d'une maniere uniforme, ou faire les pieces de la même espèce, qui les composent, parfaitement semblables; en forte qu'une de ces pièces ètant usée, cassée ou perdue, on pût aisément lui en substituer une autre, tellement identique qu'elle en remplit exactement les fonctions, ou que les effets en fussent entiérement les mêmes.
6. It will be easy to understand from these various observations, without our being obliged to add others, of what importance it would be to the army, that the locks of muskets could be manufactured in a uniform manner, or make the pieces of the same type, which compose them, perfectly similar, so that if one of these parts was worn, broken, or lost, it could easily be substituted for another so identical that it fulfilled the functions exactly, & that the effects were entirely the same.

7. On fait tous les avantages qu'on a retires de l'exacte uniformité que feu M. de Gribeauval a introduite dans toutes les parties de l'Artillerie des différens Départemens; mais il faut convenir que difficultés qu'a pu comporter l'exécution de ce projet, ne peuvent fe comparare celles que présente celui de la fabrication de platines don't toutes les pieces de la même espèce, seroient identiques, ou auroient l'avantage dont nous venons de saire mention.
7. All the advantages which have been derived from the exact uniformity which the late M. de Gribeauval has introduced into all the parts of the artillery of the different departments, but it must be admitted that the difficulties involved in the execution of this project cannot be compared with those of the manufacture of locks in which all the pieces of the same kind would be identical to have the advantage which we have just mentioned.

8. Quoi qu'il en soit, c'est cet avantage que M. Blanc entreprend de leur procurer aujourd'hui, & en quoi consiste le projet qu'il a propose au Ministre; mais ce projet est-il practicable, & la nature & la sujétion des pieces d'une platine ne s'opposent-elles pas a son exécution? C'est ce qu'il faut examiner.
8. Be that as it may, it is this advantage which M. Blanc undertakes to procure for them today by means of the project he has proposed to the Minister; but is this project practicable, & is not the nature of the parts of a lock opposed to its execution? That is what we need to examine.

9. Il est vrai que, lorsque l'on considère les pieces principals d'une platine, comme, la noix & la gachette le chien, & ce qu'elles demandent d'exactitude & de precision dans leurs dimensions, leus forms & leur position pour produire leurs effets sûrement, il semble qu'on ne peut obtenir dans l'exécution ce que se propose M. Blanc.
9. It is true that when we consider the principal pieces of a lock, such as the cock, the tumbler, & the sear, which require their precise dimensions & shape to produce their effects with certainty, it was felt that it was impossible to obtain in practice what Mr. Blanc proposed.

10. En effet, on concevra sans peine, pour peu qu'on y réfléchiffe, combienil doit être difficile, par exemple, de faire une noix qui, substituée a une autre dans une platine, se raccorde si parfairement aves la gachette, que le repos du chien soit aussi bien assure qu'auparavant, & que la détente se fasse précisement comme elle doit se faire; enfin que le chien se trouve situé sur cette nouvelle noix, exactement comme il l'étoit sur l'ancienne, en sorte qu'il parcoure le même espace, & tombe sur la batterie, au même endroit qu'auparavant, pour faire le plus grand feu.

10. In fact, it will be easy to conceive, if we reflect on it, how difficult it is, for example, to make a tumbler which, substituted for another in a lock, connects so perfectly with the sear, that the rest [position] of the cock be as well assured

as before, & that the trigger be made precisely as it ought to be done; finally, that the cock is situated on this new tumbler exactly as it was on the old, so that it traverses the same space, & falls on the frizzen in the same place as before, to be certain of sparking.

11. Cependant, quand on connoit les ressources des Arts, & les moyens qu'on emploie dans plusiers pour arrive a l'exécution de pieces entièrement semblables, on cesse de regarder le projet de M. Blanc comme impracticable; & on est dispose a croire qu'un homme habile, en employant ces moyens d'une maniere appropriée a la objet, peut parvenir a son but.

11. When, however, we know the resources of the arts, & the means employed in several of them, to arrive at the execution of entirely identical pieces, we cease to regard M. Blanc's project as impracticable; & we are disposed to believe that an able man, by employing these means in an appropriate manner may attain his objective.

12. L'Académie va en juger par l'exposé que nous allons faire de ceux dont l'Auteur se sert pour fabriquer ses platines; en forte que toutes leurs pieces, de la même espèce, aient l'identité *de situation, de forme & de dimension*, dont nous venons de parler.

12. The Academy will judge by the exposition which we are about to make of those which the Author uses to make his locks so that all their pieces of the same kind are identical in position, form, & dimension.

13. Pour ne pas charger cet exposé de details superflus, nous parlerons d'abord de cette plaque de fer oblongue, que les Arquebusiers nomment corps de platine; & sur laquelle font montées ou vissèes toutes les pieces qui, par leur union, composent ou forment rèellement ce qu'on appelle une platine.

13. In order not to load this account with superfluous details, we shall first speak of what the gunmakers call a lockplate & on which are mounted or screwed all the parts which, by their union, compose or form what is called a lock.

14. Or on peut regarder cette plaque comme une espèce de plan, sur lequel font marqués par des trous, d'une maniere invariable, toutes les places que les pièces doivent y occupier.

14. Now this lockplate may be regarded as a sort of plan, on which are holes in invariable relative positions & define the places where the parts must be fixed there.

15. On voit, en conséquence, qu'elle mérite une attention particulière, pour qu'on connoisse bien les moyens par lesquels l'Auteur parvient a y percer ces différens trous, de manière que dans tous les corps de platine, percés avec les mêmes moyens, les pièces y occupent toutes les mêmes places, & que leurs

situations respectives soient identiques, condition dont nous avons déja fait voir toute l'importance.

15. Consequently, it deserves special attention, so that the means by which the author succeeds in making these different holes will be well known, so that in all the lockplates, pierced with the same means, the components all occupy the same places, so their respective positions are identical, which is of primary importance.

16. Le corps de platine étant forge avec ses deux eminences trapezoids du côté du fût, que les Ouvriers appellant très-improprement pivots, & qui en donnant de la force a ce corps de platine, forment, avec le bassinet, ce qu'ils nomment le rempart, on donne a cette plaque le forme déterminée.

16. The lockplate being forged with its two trapezoidal eminences on the barrel side, which the workers very improperly call pivots & which form with the pan what is called the rampart, the lockplate is given its correct shape.

(mention of filing jig)

17. Pour cet effet, on la monte & on la lime sur une autre, qui est trempée, & qui a précisément la forme que la première doit avoir, etant finie.

17. For this purpose it is mounted & filed on another, which is hardened, & which has precisely the shape which the former must have when finished.

18. Cette operation faite, on passe a celle qui regarde les trous.

18. This operation done we pass on to regard the holes.

19. On atache la plaque solidement sur une matrice suffisamment épaisse, pour diriger les poincons bien perpendiculairement.

19. The plate is fixed firmly on a sufficiently thick jig to direct the punch perpendicularly.

20. Dans cette matrice sont percés tous ces trous exactement dans la position, & aux distances respectives les uns des autres, ou ils doivent être.

20. In this matrix are pierced all these holes exactly in the position, & at the respective distances from each other, as in the finished lockplate.

21. On commence par les marquer avec un poincon sur la nouvelle plaque: cette operation faite, on achève de les percer avec un autre poincon légèrement conique.

21. We start by marking them with a punch on the new plate: this operation done, we finish by piercing them with another slightly conical punch.

22. On les rend ensuite cylindriques, au moyen d'un util qu'on appelle un faux-foret, employé avec un conducteur, pour que ces trous soient bien perpendiculaires aux deux surfaces: enfin avec un alésoir, on leur donne la grosseurs déterminée, & après on taraude ceux qui doivent ètre taraudés.

22. They are then made cylindrical, by means of a tool called a false drill, used with a conductor [pilot?], so that these holes are perpendicular to both surfaces: finally, with a reamer, they are given the determined size, & then tap those which have to be tapped.

23. Par ces diverses operations, on a un corps de platine, dont la forme est toute semblable a celle qui a été arrètée, & où tous les trous se trouvent précisément dans les mèmes endroits que ceux de la plaque matrice, de facon que les pieces qu'on y placer seront respectivement dans la mème situation sur cette plaque que sur toute autre, faite avec cette mème matrice.

23. By these various operations we have a lockplate, the shape of which is very similar to that which has been determined upon & where all the holes are precisely in the same places as those of the lockplate jig, so that the parts to be placed there will be respectively in the same situation on this lockplate as on any other, made with this same jig.

24. On n'aura pas mème a craindre que l'inclinasion des trous apport un changement dans la position des pieces (ce qui pourroit arrive a cause de l'épaisseur de la plaque), puisqu'ils sont tout bien perpendiculaires a ses deux surfaces.

24. It will not even be feared that the inclination of the holes will bring about a change in the position of the parts (which might happen because of the thickness of the plate), since they are all perpendicular to its two surfaces.

25. Le corps de platine fait, il faut faire voir comment on donne a la noix & au chien les dimensions, la figure & la position qu'ils doivent avoir.

25. The body of the lockplate made, it is necessary to show how the tumbler & the cock are given the dimensions, the shape, & the position they ought to have.

26. La noix ayant été préparée, & son gros pivot arrondi, on la réduit a sa veritable épaisseur en fraisant ses deux surfaces; ensuite on la monte entre mâchoires, on pieces trempées, exactement de la mème forme que celle que doit avoir tout son contour, & on la lime dessus; mais on ne borne pas la; après l'avoir ainsi limée, on la fait passer a travers une matrice faite exprès, pour qu'il ne reste rien a desirer dans la perfection de ses crans on entailles.

26. The tumbler having been prepared with its large rounded pivot, it is reduced to its true thickness by milling its two surfaces; then it is mounted between hardened jaws of exactly the same shape as that which its entire contour must have, & it is filed on it; but we do not stop here; after having filed it in this way, it is pressed through a specially made die, so that nothing remains to be desired in the perfection of its notches.

27. Cependant ces entailles devant avoir une position déterminée par rapport au chien, on fait le carré de la noix a la faveur d'une machine qui conduit la lime de l'Ouvrier, & qui fait que les pans de ce carré sont places directement

comme ils doivent l'ètre, pour qui le chien étant perpendiculaire, ou dans la situation du repos, la gachette se trouve engage dans le cran ou l'entaille de la noix qui répond a cette position.

27. However, since these cuts must have a certain position relative to the cock, the square of the tumbler is made by a machine which guides the file of the workman, & which causes the sides of this square to be placed as they are required for the cock to be in the 'safe' position, perpendicular, when the sear is engaged in the notch of the tumbler which responds to this position.

28. L'Auteur emploie encore plusieurs autres moyens, bien imaginés, pour donner a ce carré les facons nécessaires.

28. The author employs several other means, well-conceived, to give this square its necessary orientation.

29. Mais touts ces soins seroient inutiles, si le chien n'avoit pas sur la noix exactement la position qu'il doit y avoir; aussi le trou par lequel le chien entre sur ce carré est-il percé au mouton[12], dans un moule, en sorte que monté sur cette noix, il se trouve précisément dans cette position.

29. But all this care would be useless if the cock was not fitted on the tumbler in exactly the position needed; also, the hole in the cock which fits on this square is pierced by a stamping machine, in a jig, so that when mounted on this tumbler, it is precisely in the correct position.

30. La partie do chien, tournée contre le corps do platine, ayant été préparée avec une fraise, on applique cette pièce contre un calibre très-épais, auquel on l'assujettit; & ce calibre ayant toutes les formes du chien, on le lime dessus.

30. The part of the cock which turns against the lockplate, having been prepared with a milling cutter, this piece is fixed against a very thick template which has the shape of the cock & is filed against it.

31. On emploie des soins semblables pour donner a la batterie la forme qu'elle doit avoir, ainsi que l'inclination necessaire a sa face, pour que le chien, en s'abbattant, y tombe a l'endroit le plus avantageux pour donner le feu, la faire élever & decouvrir le bassinet.

31. Similar care is employed to give the frizzen the shape it must have, as well as the inclination necessary to its face, so that the cock, by falling, strikes it in the most advantageous place to produce sparks, raise it & uncover the pan.

32. Nous avons fait connoître les moyens que M. Blanc emploie pour donner a la noix & au chien, qui sont les pieces les plus importantes de la platine, l'identité de forme & de dimensions nécessaires.

32. We have made known the means which M. Blanc employs in giving to the tumbler & the cock, which are the most important pieces of the lock, the identity of form & dimensions necessary.

33. Nous n'entreprendrons pas de parler de ceux dont il se sert, pour le bride qui maintient la noix, non plus que peur la gachette qui en règle les mouvemens.
33. We shall not undertake to speak of those which he employs, for the bridle which holds the tumbler nor for the sear, which regulates its movements.

34. On concevra sans peine qu'étant parvenu a donner cette identité aux pièces principals, il aura su la donner également aux pièces qui ne sont pas de la mème conséquence.
34. It will be understood without difficulty that, having succeeded in giving this identity to the principal pieces, he will have been able to give it also to the pieces which are not of the same consequence.

35. Nous ajouterons seulement que l'identité de toutes les pièces des platines de M. Blanc, résulte de la perfection des matrices, des moules & des mandrins qui sevent a les faire.
35. We shall only add that the identity of all the parts of M. Blanc's locks is the result of the perfection of the dies, the jigs, & the mandrels which are used in making them.

36. Or, on fait a quell degré de précision on parvient, dans différens Arts, a exécuter par ces moyens des pièces semblables.
36. Now, a certain degree of precision is obtained, in different arts, to execute by these means similar pieces.

37. Enfin il s'est tellement assure de la trempe pour qu'elle n'altère pas la forme de ces pièces, qu'il n'a point a craindre que cette opération y apporte aucun changement qui en empêche les effets.
37. Finally, he is so sure of the case hardening that it does not alter the shape of these parts, that he does not have to fear this.

38. Cependant il se présente ici une question importante.
38. However, here an important question arises.

39. En accordant qu'avec les moules, les matrices & les machines que M. Blanc a su imaginer & construire pour faire ses platines, il puisse en exécuter dont toutes les pièces, du la même espèce, soient identiques, on demandera si ces moules & ces matrices, &c. venoient a s'user, a se casser ou a se perdre, il pourroit en refaire d'autres, parfaitement semblables?
39. In granting that, with the moulds, dies, and machines which M. Blanc has been able to devise & construct to make his locks, he can execute them with all the parts of the same kind, if these jigs &c. were to wear, to break, or be lost, could he make others, perfectly alike?

40. Si les pièces qu'il fabriqueroit avec ces nouvelles matrices seroient entièrement pareilles a celles qu'il fabriquoit auparavant?

40. Would the pieces he manufactured with these new tools be exactly the same as those he had manufactured before?

41. Enfin, si, par exemple, il pourroit enjoyer dans une Manufacture un assortiment de tous les différens moules & matrices, &c. tel qu'on pût faire avec cet assortiment des platines, dont toutes les pièces de la même espèce sussent si semblables a celles des platines qu'il fabrique, qu'il n'y eût aucune différence, & qu'on pût les substituer les unes aux autres, sans craindre que les effets de ces platines en sussent deranges?
41. Finally, if, for example, he could send into a manufactory an assortment of all the different dies & jigs, & c. to make lock parts, would all the pieces of the same kind be so similar to those of the locks which he manufactures that there would be no difference, & that they could be substituted without fearing that the locks would be affected?

42. Nous croyons pouvoir répondre par l'affirmative.
42. We believe we can answer in the affirmative.

43. En effet, telle est la sûreté & la précision des moyens que M. Blanc a employés dans la construction des types, qui servent a former ses moules & ses matrices, que nous pensons qu'il est en état de pouvoir résoudre ce difficile problème d'exécution.
43. Indeed, such is the safety & precision of the means employed by M. Blanc in the construction of the types which serve to form his tools that we think he is able to solve this difficult problem of execution.

44. Il résulte de tout ce que nous venons d'exposer, que le projet de M. Blanc, de construire toutes les platines d'une maniere uniforme & de facon que toutes les pièces de la même espèce, soient identiques, est très-exécutable, par les moyens qu'il a imaginés.
44. It follows from all that we have just stated that M. Blanc's plan of constructing all the locks in a uniform manner, & in such a manner that all the pieces of the same species are identical, is very practical by the means he devised.

45. Mais quand nous disons qu'il est executable, c'est pour nous conformer a la marche que nous sommes tracée: car l'Auteur l'a déja exécuté.
45. But when we say that it is executable, it is to conform to the course we have traced: for the Author has already executed it.

46. En effet, ayant fait, il y a quelques années, des fusils pour les Officiers du ci-devant Régiment du Roi, sur un modèle particulier; il y joignit, en les envoyant, des boîtes pleines respectivement des différentes pièces d'une platine de ces fusils, pour s'en server quand il en viendroit a manquer.
46. Indeed, having made, some years ago, muskets for the Officers of the former regiment of the King, on a particular model; &, when sending them, enclosed

boxes full of the different parts of a lock of these muskets, to be used when necessary.

47. Et l'expérience fit voir que ces pièces de rechange remplacoient tellement celles auxquelles on les substituoit, que les effets en étoient les mêmes.
47. Experience has shown that these spare parts so accurately replaced those for which they were substituted, that their effects were the same.

48. Mais depuis, M. Blanc a encore perfectionné ses moyens d'exécution.
48. But since then, M. Blanc has improved his methods of execution.

49. Nous avons vu a l'Hotel des Invalides, comme beaucoup d'autres personnes, militaires & autres, prendre au hazard dans des boîtes contenant les différentes pièces d'une platine, successivement une de ces pièces, les monter sur des corps de platine, & en former des platines de fusil faisant parfaitement leurs effets.
49. We have seen at the Hotel des Invalides, like many other persons, soldiers, & others, take at random in boxes containing the different parts of a lock, successively one of these pieces, mount them on lockplates, & create perfectly effective musket locks.

50. La Compagnie vient d'avoir sous ses yeux la prevue de ce que nous venons de rapporter.
50. The Company has just had before it the anticipation of what we have just related.

51. On lui a fait voir une platine, formée des ces pièces, prise au hazard dans les boîtes a l'Hotel des Invalides, & conservée soigneu-sement par M. Tillet, & qu'une noix faite depuis par M. Blanc, & substituée a celle qui y étoit, se raccordoit si bien avec toutes les pièces de sa dépendance, que les effets étoient entièrement les mêmes qu'auparavant.
51. He was shown a lock, formed of these pieces, taken at random from the boxes at the Hotel des Invalides, & carefully preserved by M. Tillet, & a tumbler made by M. Blanc, & substituted & which so accurately connected with all the necessary parts that the lock worked entirely the same as before.

52. Il est essential d'ajouter a tout ceci, que les pièces continues dans chacune des boîtes, aux Invalides, étoient en si grand nombre, qu'on auroit pu, en les montant toutes, en former mille platines.
52. It is essential to add to all this that the component pieces in each of the boxes, at the Invalides, were in so great a number that they could have been formed into a thousand locks.

53. On voit ainsi, comme nous l'avons dit, que le projet de l'Auteur est non seulement executable, mais exécuté & avec un succès qui paroît ne rien laisser a desirer.

53. We see thus, as we have said, that the project of the Author is not only executable, but executed, & with a success which seems to leave nothing to desire.

54. Nous concluons, en conséquence, que les moyens d'exécution employés par M. Blanc pour la fabrique des armes a feu, en sorte que leurs platines soient toutes formées *de pièces parfaitement semblables* dans leurs espéces, & ains *propres a être substituées les unes aux autres,* sont trés-bien concus & imaginés; que l'effet en est assure par l'expérience, & qu'il est fort a desirer qu'on fasse toutes nos armes d'après ces moyens: enfin qu'ils méritent les éloges & l'approbation de l'Académie.

54. We conclude, therefore, that the means of execution employed by M. Blanc for the manufacture of firearms, so that their locks are all formed of separate pieces perfectly similar & capable of being substituted for each other is very well conceived & devised; that the effect is assured by experience, & that it is very desirable that we should make all our arms according to these means. Finally, they deserve the praise & the approbation of the Academy.

Fait dans l'Académie des Sciences, le Samedi, 19 Mars 1791
Signé le Roy, Laplace, Coulomb, Borda
Je certifie le present Extrait conforme a l'original & au judgement de l'Académie. A Paris, le 27 Mars, 1791.

Signé, Condorcet, Secrétaire perpétuel.

Made at the Academy of Sciences, on Saturday, 19 March, 1791
Signed: le Roy, Laplace, Coulomb, Borda
I certify the present Extract conforming to the original & judgment of the Academy, Paris, on the 27 March, 1791.

Signed, Condorcet, Permanent Secretary

Bibliography

Abbott, J. (1852). *The Armory at Springfield,* Harper's New Monthly Magazine, No. XXVI, July 1852, Vol. V, pps. 145-161.

Alder, K. (1997). *Engineering the revolution – arms and enlightenment in France, 1763 – 1815,* Princeton, New Jersey; Princeton University Press.

Ashburn, A. (1977). *The American system is born,* American Machinist, 100th anniversary issue, vol. 121, no 11, section B. New York, McGraw-Hill Inc.

Barraclough, K. (1984). *Steelmaking before Bessemer; Vol. 2. Crucible Steel,* London; The Metals Society.

Battison, E. (1984) Eli Whitney and the milling machine, *Smithsonian Journal of History,* (1), (1966, revised 1984), pps 9-34.

Benton, Lt. Col. J. G. and Capt. J. E. Greer (1878). *Fabrication of Small Arms,* Ordnance Memorandum No. 22, Washington, Government Printing Office. Reprinted 1970 by Benchmark Publishing Company Inc., Glendale, New York.

Bessemer, H. (1905). *An Autobiography,* London, Offices of 'Engineering', 1905;

Bion, N. (1709) *Traité de la construction et des principaux usages des instruments de mathématique,* [Treatise on the construction and main uses of mathematical instruments], Paris, Collombat and Boudot fils, 1709, Pl. Dixieme, p 144.

Blackmore, H. (1961). *British Military Firearms 1650 – 1850,* London, Herbert Jenkins.

Blanc, H. (1790). *Mémoire important sur la fabrication des armes de guerre,* A L'Assemblee Nationale, Paris, printed L M Cellot, rue des Gr.-August.

Clark, D. K. (1878). *A manual of rules, tables and data for mechanical engineers,* London, Blackie and Sons.

Cooper, C. and Lindsay, M. (1980) *Eli Whitney and the Whitney Armory,* Whitneyville, Connecticut; Eli Whitney Museum.

Cotty, H. (1806). *Mémoire sur la Fabrication des armes portatives de Guerre,* [Memorandum on the manufacture of portable weapons of war], Paris, Magimel. Reprinted by Facsimile Publisher, Delhi, n.d. but printed on demand 2017.

Dale, M. (1810). *Mémoire sur le montage du fusil,* [Memorandum on stocking, hardening and tempering the lock, inspection and acceptance of the finished weapon], Liege. (Reprint published on the initiative of Fabrique Nationale Herstal on the occasion of the inauguration of its collection of industrial archeology as part of the fourth Colloquium of Industrial Archeology. Liege 1977.)

Diderot, D. (1751). Diderot et D'Alembert. *L'Encyclopédie: Fabrique des armes escrime, Fusil de Munition*, Paris. 1751-1765.

Diderot, D. (1779) Diderot et D'Alembert, *L'Encyclopédie: Fabrique des armes escrime, Fusil de Munition*, Paris.

Diderot, D. (1784). *Recueil de Planches de Encyclopédie – Fabrique des armes escrime – Fusil de munition*, Paris, Pancoucke. Pl. 1, Fig. 4

Diderot (et D'Alembert). *L'Encyclopédie: Fabrique des armes escrime, Fusil de Munition*. (Inter-livres) [An undated reprint.]

Durfee W.F. (1894) The first systematic attempt at interchangeability in firearms, *Cassier's Magazine*, New York, 469-477.

Fiosconi, C. and Guserio, J. (1718). *Espingarda Perfeyta (The Perfect Gun)*, Western Lisbon, Antonio Pedrozo Galram.

Fiosconi, C. and Guserio, J. (1718). *Espingarda Perfeyta (The Perfect Gun)*, Western Lisbon, Antonio Pedrozo Galram. Reprint edited and translated by Rainer Daehnhardt and W. Keith Neal, London and New York, Sotheby Parke Bernet, 1974.

Fitch, C. H. (1882). *Report on the Manufacture of fire-arms and ammunition*, Extra Census Bulletin, Washington. Reprinted by American Archives Publishing, Topsfield, Massachusetts, 1970.

Gamel, J. (1826). *Описание Тульского оружейного завода с учетом исторических и технических аспектов*, [Description of the Tula Weapon Factory in regard to historical and technical aspects]. Moscow: August Semen Press.

Gamel, J. (1988). *Description of the Tula Weapon Factory in regard to historical and technical aspects*, Moscow, August Semen Press, 1826. Translated reprint, E. A. Battison (Ed.), published for the Smithsonian Institution Libraries and The National Science Foundation, Washington D.C. by Amerind Publishing Co. Pvt. Ltd., New Delhi, 1988.

Gooding, J. and Morton, J. (2009) Honoré Blanc and Interchangeable Gunlocks, *Journal of the Arms and Armour Society,* Vol. XIX No. 5, 208 – 218.

Goodman, J. D. (1866). In *The Birmingham gun trade*, in Samuel Timmins, (ed.), *Birmingham and the Midland hardware district*, originally published by Robert Hardwicke, London, 1866; reprinted by Frank Cass & Co Ltd, 1967.

Gordon, R. (1987). *Sixteenth-Century Metalworking Technology used in the Manufacture of two German Astrolabes*, Annals of Science, 44, pps. 71-84.

Gordon, R. (1988). Material Evidence of the manufacturing methods used in 'Armory Practice', *Journal of the Society for Industrial Archeology,* Volume 14, No 1, 23-35. Washington, D.C.; N.M.A.H., Smithsonian Institution.

Greener, W. (1841). *The science of gunnery*, London, Longman & Co., and Newcastle upon Tyne, F. Loraine.

Greener, W. W. (1884). *The Gun and its development*, 2nd Edn., London, Paris & New York, Cassell & Company Ltd.

Hadfield, Sir R. (1925), *Metallurgy and its influence on modern progress*, London, Chapman Hall.

Hicks, Major J. E. (1957). *U.S. Firearms: Notes on U.S. Ordnance Vol. 1.,* La Canada, California, James E. Hicks and Son.

James, H. and Jones, J. (1812) , *Method of Welding Gun Skelps, and Machinery for Turning Barrels of Fire-arms*, Patent No. 3469, 23rd January.

Kauffman, H. (1960). *The Pennsylvania-Kentucky Rifle,* Harrisburg, Pennsylvania, Stackpole Books.

Kipling, R. (1896) *The Seven Seas*, Methuen & Co.

Klein, H. (1988). *The Science of Measurement*, New York, Dover Publications Inc.

McNeil, I. (1968). *Joseph Bramah: a century of invention, 1749-1851,* New York, Augustus M. Kelley.

Marks, E. (1903). *The manufacture of iron and steel tubes*, Manchester, Technical Publishing Company.

Mirsky, J. and Nevins, A. (1952). *The world of Eli Whitney*, New York, The Macmillan Company.

Moxon, J. (1703). *Mechanick Exercises or the doctrine of handy works*, London, D. Midwinter; reprinted by Early American Industries Association, 1979.

Nash, S. (1818). *Turning Musket Barrels,* US Patent, (later numbered No. 2939 or 2939X)

Peaucelle, J-L, (2005) The concept of interchangeability and its achievement, *Managing and understanding, June, N° 80*. 58 – 75.

Pettibone, D. (1814). *Boring gun barrels,* US Patent, later numbered No. 2064X.

Petrie, Capt. M. (1865). *Equipment of Infantry*, London, Her Majesty's Stationery Office.

Réaumur, R. 1722, *L'art de convertir le fer forgé en acier, et l'art d'adoucir le fer fondu, ou de faire des ouvrages de fer fondu aussi finis que le fer forgé*, [The art of converting wrought iron into steel, and the art of softening molten iron, or making works of molten iron as finished as wrought iron], Paris, Michel Brunet.

Report (1854). *Report from the Select Committee on Small Arms*, London; House of Commons.

Roe, J. W. (1912). History of the first milling machine, *American Machinist, 1912, No 36.* 1037-8.

Roe, J. W. (1916). *English and American tool builders,* New Haven, Connecticut; Yale University Press.

Rolt, L. (1965). *Tools for the job*, London; B. T. Batsford Ltd.,

le Roy, de Borda, de la Place, and Coulomb (1791). *Rapport Fait à L'Academie Royale des Sciences D'Un Mémoire Important, De M. Blanc, sur la fabrication des Armes de guerre*, Le Samedi 19 Mars 1791. Extrait des Registres de l'Academie Royale des Sciences, Paris, Du 19 Mars 1791.

Round and Whitford, (1838). *Application of dies and presses to manufacture of gun and pistol locks.* British patent No. 7712.

Singer, C., Holmyard, E., Hall, A. and Williams, T. (1958). *History of Technology*, Vol. IV. Oxford, Clarendon Press.

Smith, M. (1977). *Harper's Ferry and the new technology: the challenge of change,* Ithaca and London; Cornell University Press.

Smithurst, P. (2019). *France, Russia and early interchangeability in firearms*. Arms and Armour, journal of Royal Armouries, Vol. 16, No. 2, pps, 165 -174.

Smithurst, P. (2000) *Artillery after the Crimea; large gunmaking becomes a science,* Leeds; Royal Armouries Yearbook, Vol 5, pps 121-132.

Taylor, G. (1920). *Files and Filing*, (translated from *La Lime* by Ch. Fremont), London; Sir Isaac Pitman & Sons Ltd.

Villons, (1716) *Machine pour la fabrique des canons de fusils*, No. 166, Paris, Recueil des machines de L'academie, T. III.

Willcox, Col. C. (1917). *A French-English Military Technical Dictionary*, Washington, D.C.; Government Printing Office. (he states in the Preface, 'In some respects the preparation of a French-English military technical dictionary is a task of great difficulty. This arises partly from the inability of the English language in all cases to supply the exact technical equivalent of a given French term, but more particularly from the extremely detailed nature of technical French nomenclature.')

Williams, D. and Harding, D. (2018). *The Enfield Lock barrel mill of 1815 and moves to replace musket barrel by grinding c1780 - c1840.* Arms and Armour, Vol. 15, No. 1, 22-57.

Woodbury, R. S. (1960) *History of the milling machine,* Cambridge, Massachusetts; M.I.T. Press.

Wyatt, J. and W., (1760) *Machinery for cutting screws*, British Patent, No. 751.

References

Chapter 1: The Manufacture of The French Model 1777 and Russian Model 1808 Muskets

1. Kipling, R., 1896.
2. Gamel, 1988, translated edition, p. 133, footnote 3.
3. Blanc, 1790.
4. Alder, 1997, p. 223.
5. Blanc, 1790, p. 6.
6. Blanc, 1790, p. 5.
7. Blanc, 1790, p. 7.
8. le Roy et al, 1791.
9. Alder, 1997, p. 240.
10. Alder, 1997, p. 247.
11. Blanc, 1790, p. 7.

Chapter 2: Barrel Manufacture

1. Fiosconi et al, 1718, translated edition, pps. 130, 131.
2. Hadfield, 1925, p. 40.
3. Fiosconi et al, 1718, translated edition, p. 79.
4. Smithurst, 2000, pps. 121-132.
5. Singer, et al, 1957.
6. Marks, 1903, p. 3.
7. Bessemer, 1905, pl. XXIV.
8. Blackmore, 1961, p. 43.
9. Fiosconi et al, 1718, translated edition, p. 113.
10. Fiosconi et al, 1718, translated edition, p. 119.
11. Smithurst, 1997, 2000.
12. Moxon, 1703.
13. Fiosconi et al, 1718, translated edition, p. 159.
14. Fiosconi et al, 1718, translated edition, p. 161.

15. Fiosconi et al, 1718, translated edition, p. 163.
16. Fiosconi et al, 1718, translated edition, p. 171.
17. Blanc, 1790, p. 6.
18. Cotty, 1806, pps. 45, 46.
19. Cotty, 1806, p. 47.
20. Cotty, 1806, p. 47, 48.
21. Gamel, 1988, translated edition, p. 53.
22. Gamel, 1988, translated edition, p. 96.
23. Gamel, 1988, translated edition, pps. 97, 98.
24. Gamel, 1988, translated edition, p. 98.
25. Gamel, 1988, translated edition, p. 97.
26. Gamel, 1988, translated edition, p. 99.
27. Clark, 1878, p. 171.
28. Gamel, 1988, translated edition, p. 100.
29. Gamel, 1988, translated edition, p. 101, footnote.
30. Greener, 1841, p. 132.
31. Greener, 1841, p. 138.
32. Greener, 1841, pps. 105-107.
33. Smith, 1977, pps. 94, 95.

Chapter 3: Finishing the Barrel Internally

1. Fiosconi et al, 1718, translated edition, p. 165.
2. Goodman, 1866, p. 394.
3. Fiosconi et al, 1718, translated edition, pps. 61, 63.
4. Fiosconi et al, 1718, translated edition, p. 67.
5. Fiosconi et al, 1718, translated edition, pps. 57-61.
6. No translation for this archaic Portuguese term has been found; from a description of their use they would appear to be similar to 'winding sticks' used in carpentry to detect any twist in a flat surface.
7. Fiosconi et al, 1718, translated edition, pps. 66, 67.
8. This seems almost certainly a mistranslation of the word *cano* which is taken to mean *reed*, not *rod*, since a similar process was used in the English gun trade where a *spill* of wood was used to pack a boring tool to make very fine cuts.
9. Fiosconi et al, 1718, translated edition, p. 69.
10. Fiosconi et al, 1718, translated edition, p. 183.
11. Fiosconi et al, 1718, translated edition, p. 185.
12. Fiosconi et al, 1718, translated edition, p. 185-193.
13. Fiosconi et al, 1718, translated edition, p. 18.
14. Fiosconi et al, 1718, translated edition, p. 209.
15. Fiosconi et al, 1718, translated edition, pps. 191, 192.
16. Cotty, 1806, pps. 49, 50.
17. Cotty, 1806, p. 49.

18. Cotty, 1806, p. 51.
19. Gamel, 1988, translated edition, p. 103.
20. Gamel, 1988, translated edition, p. 103.
21. Gamel, 1988, translated edition, p. 103, footnote.
22. Gamel, 1988, translated edition, p. 104.
23. Gamel, 1988, translated edition, p. 105.
24. Gamel, 1988, translated edition, p. 105.
25. Gamel, 1988, translated edition, p. 104, footnote.
26. Gamel, 1988, translated edition, pps. 104,105.
27. Gamel, 1988, translated edition, pps. 105,106.
28. Gamel, 1988, translated edition, p. 105, footnote.
29. Bast, in English usage, was generally taken to mean a thin strip of flexible inner bark from a Lime tree. (O.E.D.)
30. Abbott, 1852, p. 151.
31. Gamel, 1988, translated edition, pps. 105,106.
32. Gamel, 1988, translated edition, p. 106.
33. Fitch, 1882, p. 11.
34. Smith, 1977, p. 97.
35. Smith, 1977, p. 96, footnote.
36. Smith, 1977, p. 99.
37. Abbott, 1852, p. 149.
38. Fitch, 1882, p. 11.
39. Benton, 1878, pl. XVII.
40. Fitch, 1882, Fig. 7.
41. Greener, 1884, pps. 269-274.
42. Greener, 1884, p. 269.
43. Greener, 1884, p. 271.
44. Greener, 1884, p. 274.

Chapter 4: Finishing the barrel externally

1. Fiosconi et al, 1718, translated edition, p. 193.
2. Fiosconi et al, 1718, translated edition, p. 225.
3. It is described as a square file one palm (217mm) in length and two fingers (38mm) wide on each side, each end being held in the hands, as in draw-filing.
4. Klein, 1988, p. 63.
5. Fiosconi et al, 1718, translated edition, p. 199.
6. Fiosconi et al, 1718, translated edition, pps. 197, 198.
7. Fiosconi et al, 1718, translated edition, p. 207.
8. Fiosconi et al, 1718, translated edition, p. 207.
9. Fiosconi et al, 1718, translated edition, p. 209.
10. A tostão or 'testoon' was a small Portuguese silver coin approximately 18 mm diameter and 0.75 -1 mm thick.

11. Fiosconi et al, 1718, translated edition, p. 247.
12. Moxon, 1703, p. 1.
13. Villons, 1716.
14. Taylor, 1920, p. 24.
15. Bion, 1709, pl. 10, p. 144.
16. Willcox, 1917.
17. Cotty, 1806, pps. 52, 53.
18. Cotty, 1806, p. 52.
19. Alder, pps. 204, 205.
20. Wilcox, 1917.
21. Wilcox, 1917.
22. Cotty, 1806, pps. 53, 54.
23. Gamel, 1988, translated edition, p. 107.
24. Gamel, 1988, translated edition, p. 107.
25. Gamel, 1988, translated edition, p. 85, footnote.
26. Gamel, 1988, translated edition, p. 108.
27. Gamel, 1988, translated edition, p. 108.
28. Gamel, 1988, translated edition, p. 109.
29. Gamel, 1988, translated edition, p. 110.
30. Gamel, 1988, translated edition, p. 82, footnote.
31. Gamel, 1988, translated edition, p. 112.
32. Gamel, 1826, translated edition, p. 112.
33. Gamel, 1988, translated edition, p. 115.
34. Gamel, 1988, translated edition, pps. 85, 109, footnotes.
35. Gamel, 1988, translated edition, p. 115.
36. Gamel, 1988, translated edition, pps. 116, 117.
37. Gamel. 1988, translated edition, pps. 117, 118.
38. Nash, 1818.
39. Smith, 1977, p. 121.
40. Abbott, 1852, p. 8.
41. Williams et al, 2018.
42. James, et al, Patent No. 3469, 1811.
43. Greener, 1841, pps. 146-148.
44. Greener, 1884, pps. 269, 270.
45. Abbott, 1852, p. 150.

Chapter 5: Manufacture of the lock

1. Blanc, 1790, p. 15.
2. Report, 1854, p. 249.
3. Blackmore, 1962, pps. 29-31.
4. Blackmore, 1962, p. 48.
5. Fiosconi and Guserio, translated edition, p. 205.

6. Gordon, 2010.
7. Réaumur, 1722.
8. Alder, 1997, pps. 221, 222.
9. Alder, 1997, p. 240.
10. Alder, 1997, p. 247.
11. Cotty, 1806, p. 63.
12. Cotty, 1806, pps. 63, 64.
13. Cotty, 1806, p. 65-66.
14. Cotty, 1806, pps. 65-66.
15. Cotty, 1806, p. 76.
16. Alder, 1997, p. 239.
17. Alder, 1997, p. 277.
18. Alder, 1997, p. 299.
19. Alder, 1997, p. 289.
20. Alder, 1997, p. 196.
21. Cotty, 1806, p. 66.
22. Taylor, 1920, p. 25.
23. Cotty, 1806, pps. 65, 66.
24. Cotty, 1806, p. 69.
25. Alder, 1997, pps. 197, 199.
26. le Roy et. al.
27. Cotty, 1806, pps. 67, 68.
28. Alder, 1997, p. 199.
29. Alder, 1997, p. 199.
30. Cotty, 1806, pps. 67, 68.
31. Cotty, 1806, p. 83.
32. Cotty, 1806, p. 65.
33. Barraclough, 1984, Vol. 2, pps. 16, 17.
34. Barraclough, 1984, Vol. 2, p. 27.
35. Cotty, 1806, p. 66.
36. Cotty, 1806, pps. 77-79.
37. Cotty, 1806, p. 66.
38. Cotty, 1806, footnote, pps. 66, 67.
39. Cotty, 1806, p. 68.
40. Cotty, 1806, p. 68.
41. Cotty, 1806, p. 68,69.
42. Dale, 1810, pps. 35, 36.
43. Dale, 1810, p 36.
44. Cotty, 1806, pps. 75-87.
45. Dale, 1810, p. 37.
46. Dale, 1810, p. 36.
47. Dale, pps. 37, 38.
48. Bodmer, 1869.
49. Alder, 1997, p. 236.

50. Gamel, 1988, translated edition, p. 131.
51. Gamel, 1988, translated edition, p. 133.
52. Gamel, 1988, translated edition, p. 137.
53. Gamel, 1988, translated edition, p. 137.
54. Gamel, 1988, translated edition, p. 137.
55. Gamel, 1988, translated edition, p. 139.
56. Gamel, 1988, translated edition, p. 139.
57. Gamel, 1988, translated edition, p. 141.
58. Gamel, 1988, translated edition, pps. 142-144.
59. Gamel, 1988, translated edition, p. 142.
60. Gamel, 1988, translated edition, p. 143.
61. Gamel, 1988, translated edition, p. 142.
62. Gamel, 1988, translated edition, p. 143.
63. Singer et al, 1958, vol IV, p. 437.
64. Hicks, 1957, p. 18.
65. Gordon, 2010.
66. Hicks, 1957, p. 16.
67. Mirsky et al, 1952, pps. 201, 214, 224.
68. Cooper et al, 1980, pps. 12, 14.
69. Mirsky et al, 1952, p. 177.
70. McNeil, 1968, p. 46-49.
71. Rolt, 1965, pps. 84, 85.
72. Rolt, 1965, p. 89.
73. Wyatt, 1760.
74. Rolt, 1965, p. 58.
75. Roe, 1916, pps. 134-136.
76. Mirsky et al, 1952, p. 214
77. Cooper et al, 1980, p. 16.
78. Gordon, 1988.
79. Battison, 1984.
80. Durfee, 1894, p. 467.
81. Gooding et al, 2009, p. 208.
82. Mirsky et al, 1952, p. 22.
83. Alder, 1997, p. 4.
84. Mirsky et al, 1952, p. 210.
85. Kauffman, 1960, pps. 165, 166.
86. Smith, 1977, p. 86.
87. Smith, 1977, p. 89.
88. Smith, 1977, p. 91.
89. Roe, 1916, p. 134.
90. Peaucelle, 2005, p. 60.
91. Roe, 1912, pps. 1037-8.
92. Woodbury, 1960, pps. 16-19.
93. Ashburn, 1977.

94. Smith, 1977, p. 212.
95. Gordon, 1987,
96. Cotty, 1806, p. 74.
97. Alder, 1997, p. 245.
98. Alder, 1997, p. 321.
99. Cotty, 1806, p. 63 – footnote.
100. Roe, 1916, pps. 134-136.
101. Mirsky et al, 1952, p. 214.
102. Cooper et al, 1980, p. 16.
103. Gordon, 1988.

Chapter 6: Manufacture of the Stock

1. Fiosconi et al, 1718, translated edition, p. 319.
2. Cotty, 1806, pps. 14, 15.
3. Dale, 1810, p. 15.
4. Dale, 1810, p. 16.
5. Dale, 1810, p. 20.
6. Dale, 1810, p. 22.
7. Dale, 1810, p. 47.
8. Petrie, 1865, Pl. XX.
9. Gamel, 1988, translated edition, pps. 91, 92.
10. Gamel, 1988, translated edition, p. 92.
11. Gamel, 1988, translated edition, p. 155.
12. Gamel, 1988, translated edition, p. 155, footnote 1.